Quantum Certainty Illustrated

A Mathematics of Natural and Sustainable Human
History Depicted with Art

Pravir Malik, Ph.D.

Illustrations: Narendra Joshi, Ph.D.

Copyright © 2017, © 2023 Pravir Malik
All rights reserved.

ISBN-13: 979-8-9862443-5-8

To Sri Aurobindo

AUTHOR'S NOTE — 7

SECTION 1: LIGHT, A BIG BANG, THE SEED-EQUATION, & QUANTA — 10

CHAPTER 1.1: LIGHT@C, QUANTA, AND THE REALITY IT CREATES — 12
CHAPTER 1.2: LIGHT @INFINITY — 16
CHAPTER 1.3: THE BIG BANG & THE SEED-EQUATION — 18
CHAPTER 1.4: WHAT IS LIGHT & WHAT ARE QUANTA? — 21
CHAPTER 1.5: STRUCTURED TIME, STRUCTURED SPACE, AND THEIR RELATION TO QUANTA — 25
CHAPTER 1.6: THE QUANTIZATION OF SPACE, TIME, MATTER, AND GRAVITY — 27

SECTION 2: THE MATHEMATICAL FOUNDATION TO INTERPRET QUANTUM PHENOMENA — 30

CHAPTER 2.1: LIGHT-MATRIX – THE SEED-EQUATION — 31
NATURE OF LIGHT AT U — 31
NATURE OF LIGHT AT ∞ — 35
TRANSFORMATION FROM ∞ TO U — 38
CHAPTER 2.2: NATURE OF A POINT-INSTANT IN A SYSTEM — 43
CHAPTER 2.3: ARCHITECTURAL FORCES — 51
CHAPTER 2.4: UNIQUENESS OF ORGANIZATIONS — 53
CHAPTER 2.5: EMERGENCE OF UNIQUENESS — 58
CHAPTER 2.6: INHERENT DYNAMICS OF ANY SYSTEM — 64
CHAPTER 2.7: QUALIFIED DETERMINISM — 72
CHAPTER 2.8: EQUATION FOR LIGHT-SPACE-TIME EMERGENCE — 84

SECTION 3: INTERPRETATION OF QUANTUM PHENOMENA — 89

CHAPTER 3.1: INTERPRETING QUANTUM PHENOMENON – A CORE HYPOTHESIS ... 90
CHAPTER 3.2: SPEED OF LIGHT AND QUANTA ... 95
CHAPTER 3.3: INTERPRETING SCHRODINGER'S EQUATION ... 100
CHAPTER 3.4: INTERPRETING HEISENBERG'S UNCERTAINTY PRINCIPLE ... 105
CHAPTER 3.5: INTERPRETING OTHER QUANTUM PHENOMENA ... 110
DUAL WAVE-PARTICLE NATURE ... 110
INDEPENDENT STATES AS SPECIFIED BY SUPERPOSITION ... 111
QUANTUM TUNNELING ... 111
CANCELING OUT OF QUANTUM DYNAMICS ... 115
TRAVELING FASTER THAN THE SPEED OF LIGHT ... 116
ENTANGLEMENT ... 118
GOING BACKWARD IN TIME ... 119
SUMMARY ... 121
CHAPTER 3.6: A DEEPER LOOK AT QUANTIZATION OF SPACE, TIME, MATTER, AND GRAVITY ... 123
THE STRUCTURE OF SPACE ... 127
THE STRUCTURE OF TIME ... 127
ENERGY & MATTER ... 128
GRAVITY ... 129

SECTION 4: QUANTUM CERTAINTY & NATURAL HISTORY ... 131

CHAPTER 4.1: LIVING CELLS AS THE FULCRUM OF NATURAL HISTORY ... 133
NUCLEIC ACIDS ... 135
PROTEINS ... 137
LIPIDS ... 140
POLYSACCHARIDES ... 143
CHAPTER 4.2: QUANTUM-CERTAINTY AND THE ADAPTABILITIES OF THE CELL ... 146

SECTION 5: QUANTUM-CERTAINTY & SUSTAINABLE HUMAN HISTORY ... 155

CHAPTER 5.1: SENSATIONS, URGES, FEELINGS, THOUGHTS	**156**
SENSATIONS	156
URGES, DESIRES & WILLS	157
FEELINGS & EMOTIONS	158
THOUGHTS	159
CHAPTER 5.2: BASIS FOR A SUSTAINABLE GLOBAL CIVILIZATION	**161**
CHAPTER 5.3: QUANTUM-CERTAINTY & SUSTAINABLE HUMAN HISTORY	**167**

APPENDIX: A BRIEF EXPLANATION OF THE ILLUSTRATIONS IN THIS BOOK	**179**

RELEVANT BACKGROUND AND FOLLOW-UP INFORMATION	**236**
THE AUTHOR'S EARLY BOOKS	236
THE FRACTAL SERIES	236
THE COSMOLOGY OF LIGHT SERIES	236
THE APPLICATION OF COSMOLOGY OF LIGHT SERIES	237
THE ARTISTIC INTERPRETATION OF COSMOLOGY OF LIGHT SERIES	237
NOTE ON GENESIS OF BOOKS	238
ABOUT THE AUTHOR	241
ABOUT THE ILLUSTRATOR	242
SELECTED AUTHOR ONLINE PRESENCE	245

Author's Note

The world of quantum physics has been described as weird or absurd. In it are found phenomena such as dual wave-particle nature, superposition, quantum tunneling, entanglement, going backward in time, quantum teleportation, quantum uncertainty, amongst others, that appear very contrary to phenomena that we are used to in the visible day-to-day world.

This book presents a mathematical model, generated from properties of light, which proposes a context for interpreting quantum phenomena. Essentially, quantum phenomena are positioned as being at the interface between realities set up by different speeds of light.

As per the mathematical model presented here the quantum weirdness is not really weird and is a natural outcome of considering reality at the interface of different light-based worlds. While it is perhaps easier to have become aware of such phenomena at the micro-level because of the experimental rigor and attention to detail that has accompanied the study of quantum particles, it is proposed that such weirdness exists also at the macro-level. The weirdness is positioned as appearing "weird" primarily because models of reality are force-fitted into a single layer of existence, when in fact it must be a complex of multiple interacting layers as suggested by the possibility of simultaneous and even interpenetrating realities so set up by multiple speeds of light.

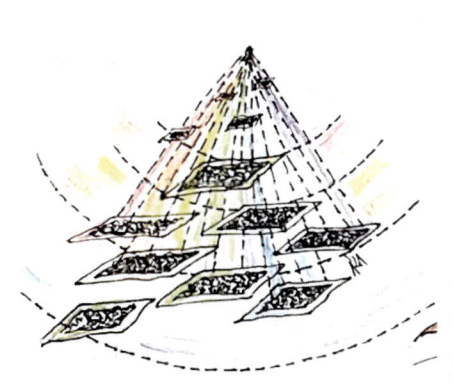

Hence this book will also explore the conditions is which similar macro-level weirdness may be more easily observed or even created. Namely, in a culminating movement, this book will hypothesize that history, whether natural or human, is the result of sustained cohesive moments or activation-states, when the multiple-layer structure accessed through quanta aligns so as to create certainty, or more precisely quantum-certainty, in outcome. In this point of view such direction-setting dynamics has to be at the quantum-level because it is only through a process of coordinated micro-level space, time, energy, and gravity quantization that the very fabric of existence and possibility can change.

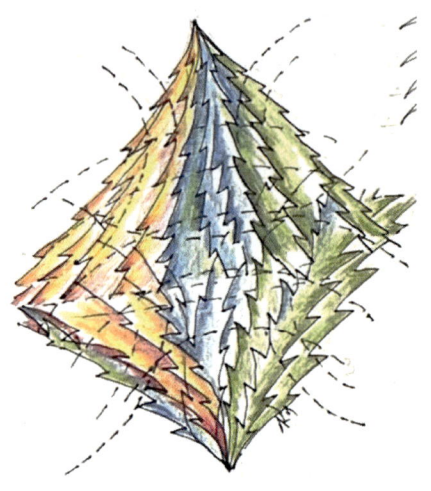

This book therefore proposes a process to create sustainable history that involves entering into an activation-state. Such an activation-state opens a quantum-window into layers of reality formed by light moving at different speeds. Collective meta-functions, whether of plants, animals, or humans exist in these layers and it is the enhancing or creation of such a meta-function that sets into motion a precise process of space, time, energy, and gravity quantization, by which the nature of reality can be changed. As such there is no uncertainty at these decisive moments of history, but a quantum-certainty that is worked out by the mathematical possibilities present at the time, and as elaborated by the mathematical model presented here.

Note that this book is the sixth in the *Artistic Interpretation of a Cosmology of Light* book series, and the sixteenth book to do with a cosmology of light and is accompanied by insightful pencil drawings and other art totaling 96 illustrations in all. The cover illustration depicts the inevitable precipitation and growth of diverse forces set into motion once deep yearning activates quantum certainty. The Rudra aspect is there to clear the ground so that space-time-energy-gravity quantization activated by quantum certainty, becomes fruitful. The Shiva aspect is there to continue to maintain some order until the next level of equilibrium is established. My hope is that the art provides an avenue for the reader to enter mode deeply into the mathematical model of light, and in particular of quantum certainty, proposed in this book.

Pravir Malik,
San Francisco

SECTION 1: LIGHT, A BIG BANG, THE SEED-EQUATION, & QUANTA

This section explores a theory for the creation of quanta related to the speed of light. Initially quanta and a seed-equation are created in the big bang. Quanta are intimately associated with the seed-equation and acts as a doorway, as it were, into the multi-layered structure represented by the seed-equation.

- Light @c, Quanta, & The Reality it Creates
- Light @Infinite
- The Big Bang & The Seed-Equation
- What is Light & What are Quanta?

- Structured Time, Structured Space, and Their Relation to Quanta
- Quantization of Space, Time, Matter, and Gravity

Chapter 1.1: Light@c, Quanta, and The Reality It Creates

Everything is made from light. But how can this be. How can the dense material objects that make up our world – the stones, cars, buildings – be made from light? How can our thoughts, our emotions, our cells and bodies be made from light? How can all the animals and plants and clouds be made from light? Here is an exploration on how this may be possible.

To begin this exploration think for a minute as to how fast light is traveling. If we turn on a light switch in a room it immediately gets filled with light. If we turn on a light torch on a dark night it

12

immediately lights up a path through the darkness. And yet the light from the sun takes about eight minutes to reach the earth. So while light is moving incredibly fast it is still not moving at the infinite speed we may think it is moving at.

Now, it is because the light is not traveling at an infinite speed that quanta and subsequently the world of matter can appear. All matter is made from atoms. So let us enter into the world of a hypothetical atom in the process of creation to explore how this might be. Imagine that some form of light originates from what will become the nucleus within an atom. By virtue of its finite speed that light would take some time, however small, to travel any minute distance within the forming atom. During that fraction of a time there is a build up of energy and it is this build up of energy that forms a packet or quanta, as it were, that allows matter to form. In the absence of such a build-up potential-matter would simply disperse. Hence, it is this build up on energy that variously expresses itself as quarks and leptons and bosons and subsequently atoms - the very building blocks of everything that is.

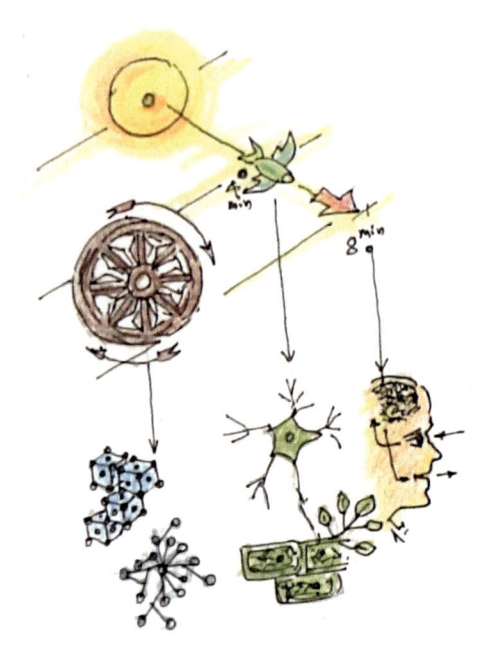

So, in this view, it is because of the finite speed of light that matter is created. That is not to say that matter will be created wherever light is traveling at a finite speed, but the possibility exists, so long as other conditions are also fulfilled. But let us leave aside the other conditions and the exact process of such a creation for now.

But what else is implicit or made possible because of the finite speed of light? Well, imagine traveling on a ray of light from the sun to the earth. Imagine that you are in minute 4 of the approximately 8 minute journey. As you look back you will see that 4 minutes in the past you were at the sun. 4 minutes in the future you will be at the earth.

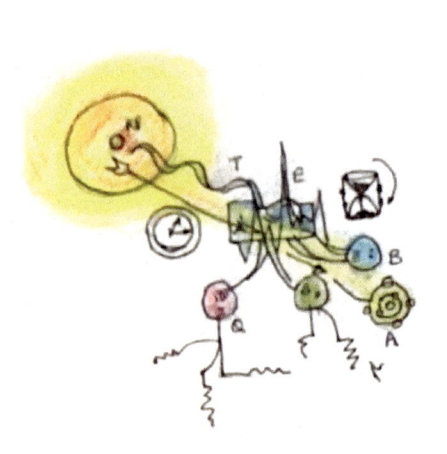

14

And in the present moment you are somewhere between the sun and the earth. So this limited speed of light already creates the concept of time and specifically of the past, the present, and the future.

So, four incredibly fundamental things are created because of the finite speed of light: quanta and therefore matter, the past, the present, and the future. It could be said, therefore, that this matter-based time-bounded universe is a result of the finite speed of light.

Chapter 1.2: Light @Infinity

But what if light can be experienced at other speeds? What if light can travel at other finite or an infinite speed beyond what we experience it at, which by the way is an astronomical 186,000 miles per second in vacuum.

Let us imagine for a minute that light can travel infinitely fast. Think about a big area or volume with a light source at the center. Now since the light travels infinitely fast it will fill up the entire volume instantly. This will be true no matter how large the volume is – it could be the entire solar system, an entire galaxy, or an entire universe. So that light is going to be instantaneously present everywhere – it is going to be omnipresent.

Now, since the light is already present everywhere in whatever volume, that is, since that light has already filled up the entire volume, there is nothing else that can similarly arise there that is not of the nature of light. Even if something else were to arise, being surrounded by light it would eventually succumb to that light. So the light is all-powerful or omnipotent within that volume.

Further, since the light exists simultaneously everywhere in that space and has a complete knowledge of itself, it therefore has a complete knowledge of that space or of anything that can arise in that space. So it is all-knowing or omniscient in that space.

Finally, the light connects everything together instantaneously and holds these connections and the things connected in its nature, so it is all-nurturing or omninurturing within that space.

So, as a result of the infinite speed of light, light appears to have properties of omnipresence, omnipotence, omniscience, and omninurturance.

Chapter 1.3: The Big Bang & The Seed-Equation

We hear about the big bang as the start of the universe. In this big bang matter is created. But in the examination presented here the creation of matter is nothing other than the result of light traveling at the finite speed, and in this universe at 186,000 miles per second in vacuum. So we can say that the big bang, the apparent start of the universe, is the result of a slowing down of light from an infinite speed to some finite speed.

When light slows down then energy accumulates in packets or quanta and this results in what was inexpressible being able to express itself as matter. This can be thought of in terms of an incredibly rapidly moving stream of water. If the water is traveling so fast then no boundary will be able to contain it and the energy will be continuous over the length of the stream. If it is traveling slower though, then the water will be able to be held by boundaries along the length of the stream. The energy in this case will be discontinuous and will appear in "packets".

In this process of slowing down the implicit omnipresent-omnipotent-omniscient-omninurturing nature of light becomes or transforms into an explicit or emergent matter-past-present-future nature of light. This process of transformation creates a seed-equation that is present and synonymous with the big bang. The dynamics of a resulting universe are contained in the dynamics of such a seed-equation. The seed-equation will be explored in more detail in the section on mathematical foundations.

We have arisen in the field of light that has a finite speed. It is difficult therefore to feel the reality of the omnipresent-omnipotent-omniscient-omninurturing light. We are attuned to perceiving in the material world that has resulted due to the finite speed of light. But if we could step back into the fullness of light in its pure state at infinity, from there we would likely see that there can be different universes created as a result of light selectively slowing down to different levels. The slow down to a different level will create a particular kind of universe.

Note that there have been experiments to slow down light so that it practically moves at a snail's pace. This slowing down has to be put into context. Even if light were to slow down to 1 mile per second, say, from 186,000 miles per second, keep in mind that it is possible that this change in speed is likely only a miniscule fraction of the change in speed from light traveling infinitely fast to light traveling at c, that is, 186,000 miles per second. So in effect what could be said is that when light slows down, the range or band that it slows down to, deviates only slightly from its slower speed in vacuum.

In this view the big bang is not an original creative event. It is likely a recurring creative event that allows something innate in light to express itself in time. But what is in light? What is in the omnipresent-omniscient-omnipotent-omninurturing nature of light? Everything we see around us, and an infinite more that we do not see.

Further, quanta being an accumulation of energy or possibility surfacing in the very process of light traveling at c, suggests the unique place it holds in a creation and in entering into anterior or contextual worlds of Light. This will be explored more fully subsequently.

Chapter 1.4: What is Light & What are Quanta?

When we think of light traveling at the speed of c, there are properties in the reality that emerge as a result of this. These properties as we have examined are of the nature of matter, the past, the present, and the future. These properties are the result of light traveling at c and so can be thought of as emerging from light.

But what do these properties actually mean? And further, how do these properties relate to the apparently quite different properties of omnipresence, omnipotent, omniscience, omninurturance that can appear in a reality where the speed of light is infinite?

Let us consider first the properties of light that emerge when light is traveling at c: the past, the present, the future, and matter.

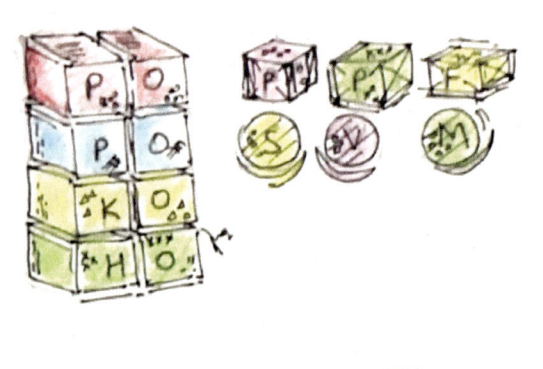

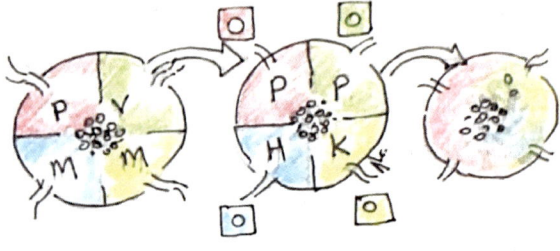

What is the past? It is the perceivable result of all the work and effort that has taken place so far. It is the foundation upon which the present and future will be built. It represents a status quo, a stability, and even a rigidity, and given that it is the result of the long play of time, it will not easily be persuaded to become another thing. It can be thought of as that which the eye can see when it looks around it. There is "physicality" to what the eye can see and so the essence of the past is a kind of physical-ness. So, ingrained in light, is this ability to project or create physical-ness.

What is the present? It is the tremendous play of forces of all kinds to express themselves here and now. There is "vitality" that is present in this play and often it is the most energetic or forceful of the forces that will win out, as opposed to the most insightful or thoughtful. All the tremendous possibility of the future is seeking for expression now and so this essential vitality can also be thought of as a projection or possibility implicit in light.

22

What is the future? It is the inevitability of what will manifest. The great thoughts, the great ideas, the purpose, the possibilities will sooner or later express themselves in what we call the future. And the essence of this is thoughtfulness or a curiosity or a purpose that we can summarize as an essential "mentality". So embedded in light is this ability to project mentality.

And what is matter? It is the myriad crystallization into apparent diversity of the one essential reality of light, to allow for a play between these different sides or possibilities in an increasingly harmonious interaction. So its essence is "harmony", and this too can be thought of as an essential property projected from or implicit in light.

But what about when light is considered to move at an infinite speed? Then the properties that become apparent, as already explored, are those of omnipresence, omnipotence, omniscience, and omninurturance. But omnipresence has physical-ness to it, omnipotence has vitality to it, omniscience a mentality to it, and omninurturance a harmony to it. It may be said that physical-ness comes from omnipresence, vitality comes from omnipotence, mentality comes from omniscience, and matter comes from omninurturance.

So whether light is traveling at an infinite speed or the speed we know as c, there is something about the properties it projects, that in essence is the same. So let us refer to these essential properties made apparent through the worlds that are created, as Presence (from omnipresence), Power (from omnipotence), Knowledge (from omniscience), and Harmony (from omninurturance).

Further, quanta can be thought of as existing on that very border of the world or reality created due to Light traveling at c, and realities created as Light travels faster, which at its limit is an infinite speed. So quanta are

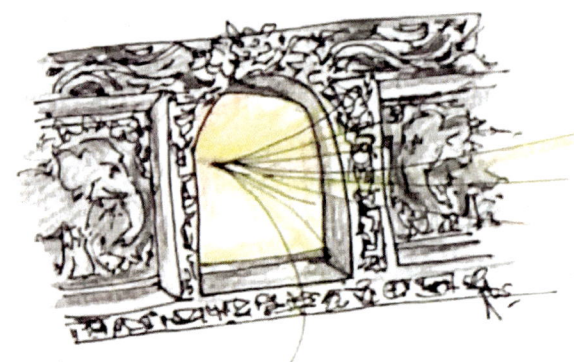

somehow a doorway or an interface into worlds of Light, and a doorway by which possibilities in deeper worlds of Light can express themselves here in this material realm.

Chapter 1.5: Structured Time, Structured Space, and Their Relation To Quanta

In considering the worlds that are created due to the way light moves or the play of light, we see properties that are projected because of it. In the finite world, the world that results from light traveling at the speed c, there is, relatively, smallness we can grasp and on which we can build. In the infinite world that results from light traveling at an infinite speed there is a vastness and fullness that is more difficult to grasp.

The notions of time and space are something entirely different in both worlds, and it could be said that Space allows the full play of everything meant by Power, Knowledge, Harmony, and Presence to be seeded in it, and that Time allows that seeding to flower into fuller forms with its passage.

In other words both space and time are not just abstract concepts but are essentially highly structured to allow physical-ness, vitality, mentality, and matter to become Presence, Power, Knowledge, and Harmony.

Space allows all the possibilities present in Presence, Power, Knowledge, and Harmony and seeded in vast diversity, to evolve into more fullness through the time stages of physical-ness, vitality, mentality, and evolving matter.

Such a creation of space and time is synonymous with the creation of quanta. Quanta become the means for the possibilities inherent in the anterior worlds of Presence, Power, Knowledge, and Harmony to express themselves in a structured space and time. Quanta are therefore a passage into deeper worlds of Light, and a means for possibilities in these deeper realms to express themselves materially.

But further, in this view it may also be proposed that space and time being so structured by the four properties of Light, need also to be quantized. That is, space and time must be experienced as quanta as well.

Chapter 1.6: The Quantization of Space, Time, Matter, and Gravity

Chapter 1.1 proposed that quanta is the result of the slowing down of light, and as such becomes the basis for material expression. In other words, for matter to express itself requires quanta. But further, in Chapter 1.5 it was also proposed that space, consisting of seeds of the properties of light, would also require quanta to express itself. Philosophically this has to be in a world where light is traveling at a fraction of its possible speed. Further, it was also proposed that time is highly structured, expressing the growth of the seeds in space through definite phases. As such growth through such phases can also be thought of as happening due to quantization that so allows phases to express themselves.

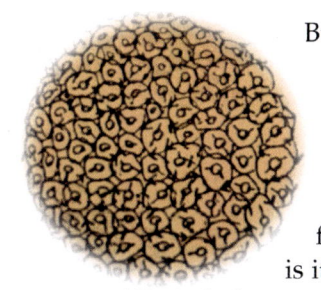

But if we take a deeper look at space, time, and matter in light of the properties of Light, it can be deduced that space, consisting of a vast array of seeds derived from the properties of Light is itself an expression of Light's property of Knowledge.

Time, bringing forth the meaning contained in the seeds, regardless of circumstance, and even being opposed by circumstance, can be thought of as Light's property of Power.

Matter itself, being a container in which space and time can allow deeper properties of Light to become materially tangible, must be an expression of Light's property of Presence.

But it is also known from Einstein's General Theory of Relativity that gravity is associated with mass and space, in that it is none other than a mass's instruction telling space how to curve, and again is nothing else that space's instruction telling mass how to move through it. As such, where mass and space exist, there gravity has to exist as well. Hence it may be inferred that gravity is none other than an expression of Light's property of Harmony, which fixes the collective relationship between object and object.

But if as proposed matter, space, and time all need to be quantized in order to express themselves, then this must also be true of gravity.

The subsequent section on Mathematical Foundations will begin to explore this in greater detail.

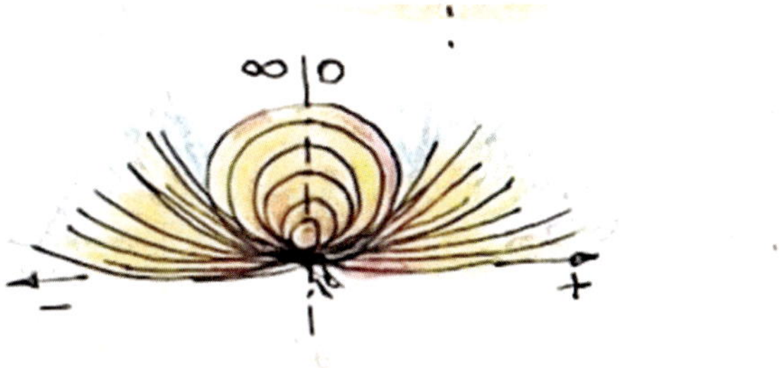

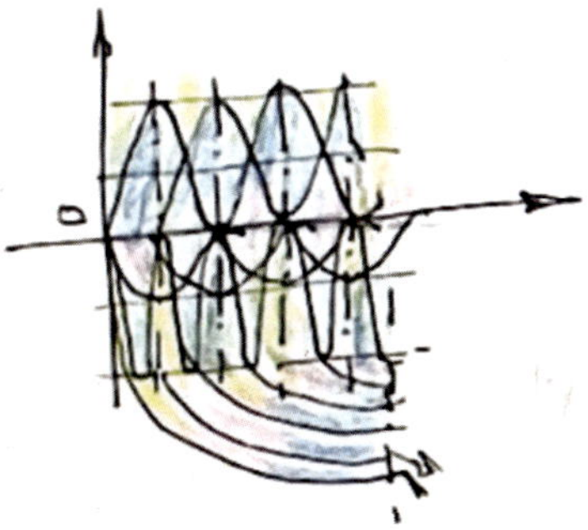

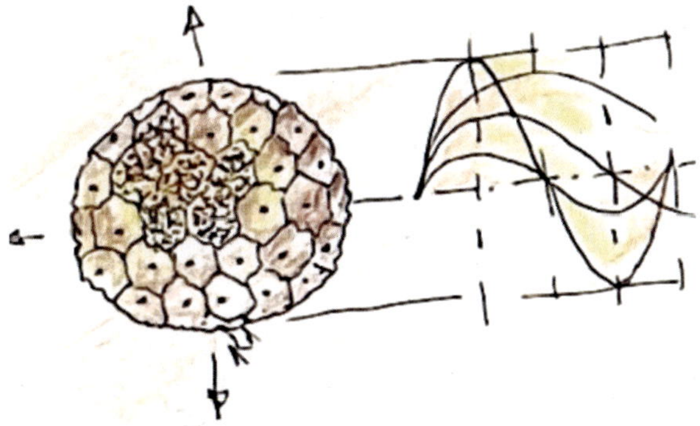

SECTION 2: THE MATHEMATICAL FOUNDATION TO INTERPRET QUANTUM PHENOMENA

This section explores the mathematics for the Light-Matrix/Seed-Equation and the dynamics of each of the layers. Since quanta are the doorway to anterior layers of Light and the quantum-level in this view is also subject to dynamics that may occur across the layers of light, the foundation presented here will allow a different interpretation of quantum phenomena.

This section will specifically explore the following foundational mathematics:
- Light-Matrix – The Seed-Equation
- Nature of Point-Instant in a System
- Architectural Forces
- Uniqueness of Organizations
- Emergence of Uniqueness
- Inherent Dynamics of Any System
- Qualified Determinism
- Equation for Light-Space-Time Emergence

Chapter 2.1: Light-Matrix – The Seed-Equation

This chapter will explore the Light-Matrix, a mathematical expression of the codification in Light formed at the time of the Big Bang. This Light-Matrix can be thought of as a seed-equation that provides insight into dynamics of the universe. The Light-Matrix codifies dynamics that may be experienced at the quantum border of realities created through Light traveling at different speeds, and further in the worlds that the quantum veil or window provide access to.

Nature of Light at U

As laid out in Section 1, it is perhaps fair to say that the speed of light has significant implications on the experienced nature of reality. The finiteness, c, at 186,000 miles per second in a vacuum (Perkowitz, 2011), creates an upper bound to the speed with which any object may travel also implying that objective reality will be experienced as a past, a present, a future from the point of view of that object (Einstein, 1995). These characteristics – a past, a present, a future – are implicit in the nature of light and become part of objective reality because of the speed of light. So the way we experience time seems definitely to be determined by c.

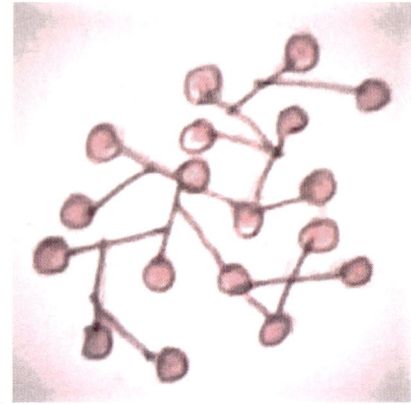

Further, c also creates a lower bound when inverted (1/c), being proportional, or arguably even determining Planck's constant, h, that pegs the minimum amount of energy or quanta required for expression at the sub-atomic level. Planck's constant, h, pegging the amount of

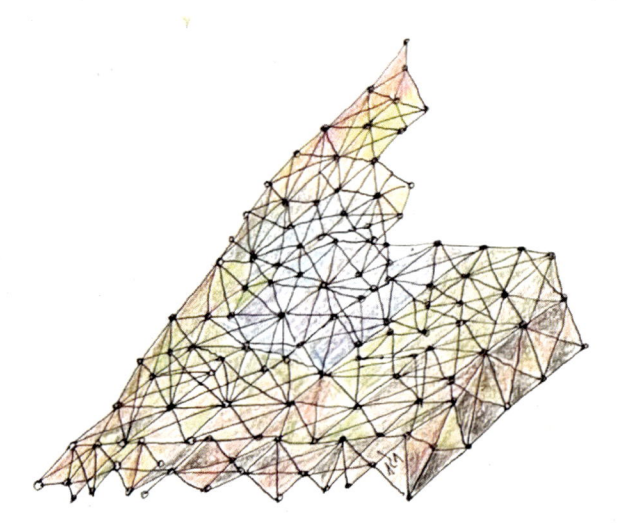

32

energy required for expression, therefore may allow matter to form (Lorentz, 1925) and for the reality of nature with a past, present, and future, to also be experienced as a phenomena of connection between seemingly independent islands of matter. This characteristic of 'connection' is therefore also proposed to be implicit in the nature of light and becomes part of objective reality because of the speed of light.

But as explored in Chapter 1.4, a 'past' can also be viewed as established reality as defined by what the eye or other lenses of perception can see. It is the perceivable result of all the work and effort that has taken place so far. It is the foundation upon which the present and future will be built. It represents a status quo, a stability, and even a rigidity, and given that it is the result of the long play of time, it will not easily be persuaded to become another thing. It can be thought of as that which the eye can see when it looks around it. There is "physicality" to what the eye can see and so the essence of the past is a kind of physical-ness. So, ingrained in light, as previously discussed, is this ability to project or create physical-ness.

The 'present' is the tremendous play of forces of all kinds to express themselves here and now. There is "vitality" that is present in this play and often it is the most energetic or forceful of the forces that will win out, as opposed to the most insightful or thoughtful. All the tremendous possibility of the future is seeking for expression now and so this essential vitality can also be thought of as a projection or possibility implicit in light.

The 'future' is the inevitability of what will manifest. The great thoughts, the great ideas, the purpose, the possibilities will sooner or later express themselves in what we call the future. And the essence of this is thoughtfulness or a curiosity or a purpose that we can summarize as an essential "mentality". So embedded in light is this ability to project mentality.

These implicit characteristics of the nature of light as experienced at the layer of reality so set up by a finite speed of light may hence be summarized by Equation (2.1.1), where c_U refers to the speed of light of 186,000 miles per second, that has created the perceived nature of reality, U, as expressed in Equation 2.1.1:

c_U: [Physical, Vital, Mental, Connection]

Eq. 2.1.1: Implicit Characteristics of the Nature of Light at U

Nature of Light at ∞

Exploring further, it is known however that at quantum levels the nature of reality is characterized by wave-particle duality. Light itself (Feynman, 1985) and matter (De Broglie, 1929) may be experienced as both particles and waves. But for matter to be experienced as waves implies that 'h' must have become a fraction of itself, $h_{fraction}$, to allow the concentration or possibility of quanta to have dispersed into wave-form. This further implies that c must have become greater than itself, c_N, such that the inequality specified by Equation 2.1.2 holds:

$$c_N > c_U$$

Eq. 2.1.2: Layer N, Layer U Inequality of Speed of Light

Note that what is implied here is that just as there is a nature of reality specified by U that is the result of the speed of light being 186,000 miles per second, so too there is another nature of reality specified by N that is the result of a speed of light greater than 186,000 miles per second.

This is akin to recent developments in physics with the notion of property spaces being separate from but influencing physical space as explored by Nobel Physicist Frank Wilczek (Wilczek, 2016). But further in "Slow Light" Perkowitz's recent treatment of today's breakthroughs in the science of light (Perkowitz, 2011) he states: "Although relativity implies that it's impossible to accelerate an object to the speed of light, the theory may not disallow particles already moving at speed c or greater. In the 1960's, Olexa-Myron P. Bilaniuk of Swarthmore College and E.C. George Sudarshan at Syracuse University began considering how to fit what they called "metaparticles" with speeds greater than c into the relativistic scheme. The approach was extended in 1967 by Gerald Feinberg (Feinberg, 1970), or Rockefeller and Columbia Universities, in his theoretical paper "Possibility of Faster-Than-Light-Particles," Feinberg also introduced the wonderful name "tachyons" for these hypothetical particles, from the Greek word "tachys" meaning swift." Perkowitz goes on to say how a flurry of papers have continued to appear about tachyons.

So light traveling at c_N seems to be possible. Current instrumentation, experience, and normal modes of thinking though, having developed as a bi-product of the characteristics so created in the layer of reality U may be inadequate to access N without appropriate modification. The notion of wave-particle duality already challenges the notion of normal thinking perhaps because wave-like phenomena could be a function of faster than c motion, and particle-like phenomena a function of less than or equal to c motion. That these may be happening simultaneously is reinforced by principles such as complementarity in which experimental observation may allow measurement of one or another but not of both (Whitaker, 2006).

But then taking this trend of a possible increase in the speed of light to its limit, this will result in a speed of light of infinite miles per second. The question is, what is the nature of reality when light is traveling at infinite miles per second? As explored in Section 1, in any space-time continuum light originating at any point will instantaneously have arrived at every other point. Hence light will have a full and immediate *presence* in that space-time continuum. Further, that light will *know* everything that is happening in that space-time completely instantaneously – that is know what is emerging, what is changing, what is diminishing, what may be connected to what, and so on - or have a quality of *knowledge*. It will connect every object in that space-time completely and therefore have a quality of connection or *harmony*. Finally nothing will be able to resist it or set up a separate reality that excludes it and hence it will have a quality of *power*.

These implicit characteristics of the nature of light as experienced at the layer of reality so set up by an infinite speed of light may hence be summarized by Equation 2.1.3, where c_∞ refers to the speed of light of ∞ miles per second, that has created the perceived nature of reality, ∞:

c_∞: $[Presence, Power, Knowledge, Harmony]$

Eq. 2.1.3: Implicit Characteristics of Nature of Light at ∞ Speed

Transformation from ∞ to U

But by (2.1.3) it is reinforced that 'physical' is related to presence, 'vital' is related to power, 'mental' is related to knowledge, and 'connection' is related to harmony.

The question then, is how do these apparent qualities at ∞ precipitate or become the physical-vital-mental-connection based diversity experienced at U? This may be achieved through the intervention or action of a couple of mathematical transformations acting on the implicit characteristics of nature of light at ∞ speed as summarized by (2.1.3).

First, the essential characteristics of Presence, Power, Knowledge, Harmony that it is posited exist at every point-instant by virtue of the ubiquity of light at ∞ will need to be expressed as sets with up to infinite elements. Second, elements in these sets will need to combine together in potentially infinite ways to create a myriad of seeds or signatures that then become the source of the immense diversity experienced at U. This suggests that all that is seen and experienced at U may be nothing other than 'information' or 'content' of light and as such that there are fundamental mathematical symmetries at play where everything at U is essentially the same thing that exists at ∞.

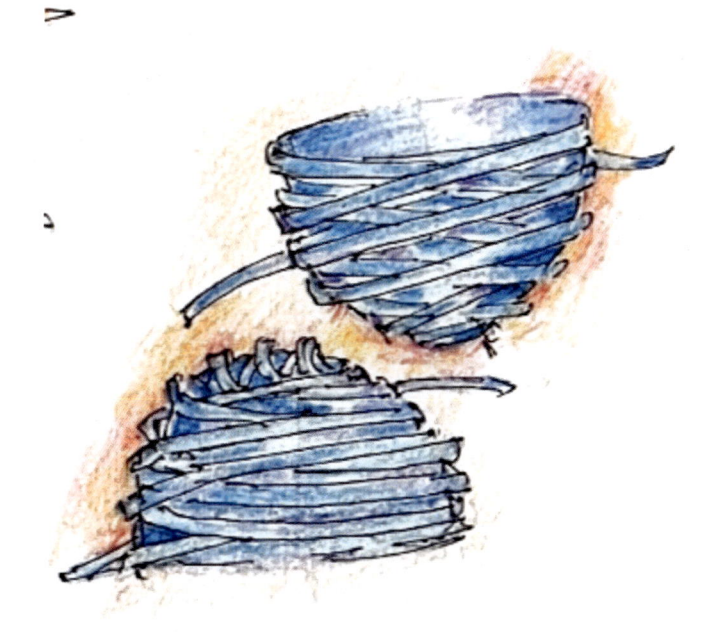

Assuming that the first transformation occurs at a layer of reality K where the speed of light is c_K, such that $c_U < c_K < c_\infty$, this may be expressed by Equation 2.1.4:

$c_K : [S_{Pr}, S_{Po}, S_K, S_H]$

Eq. 2.1.4: The First Transformation at Layer K

S_{Pr} signifies 'Set of Presence', S_{Po} signifies 'Set of Power', S_K signifies 'Set of Knowledge', S_H signifies 'Set of Harmony/Nurturing'.

Assuming that the second transformation occurs at a layer of reality N where the speed of light is c_N, such that $c_U < c_N < c_K < c_\infty$, this may be expressed by Equation 2.1.5:

$c_N : f(S_{Pr} \times S_{Po} \times S_K \times S_H)$

Eq. 2.1.5: The Second Transformation at Layer N

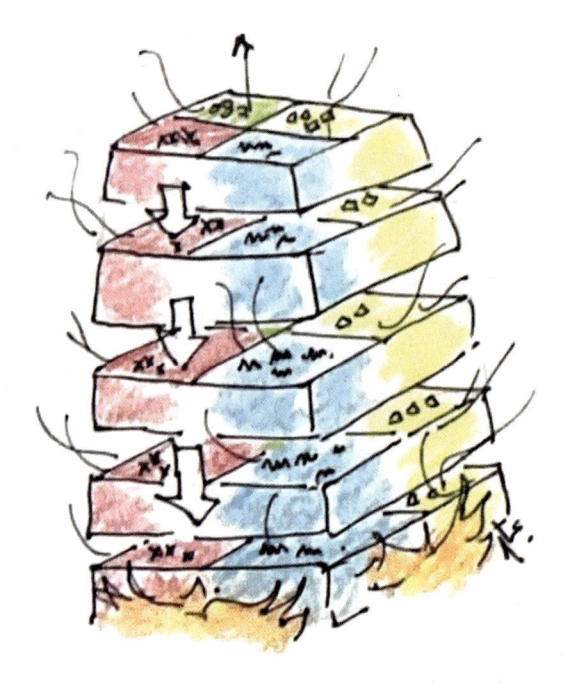

The unique seeds are therefore a function, f, of some unique combination of the elements in the four sets S_{Pr}, S_{Po}, S_K, S_H.

The relationship between the layers of light may be modeled by the following matrix in Equation 2.1.6:

$$Light_{Matrix} = \begin{vmatrix} c_\infty: [Pr, Po, K, H] \\ (\downarrow\ R_{C_K} = f(R_{C_\infty})) \\ c_K: [S_{Pr}, S_{Po}, S_K, S_H] \\ (\downarrow\ R_{C_N} = f(R_{C_K})) \\ c_N: f(S_{Pr} \times S_{Po} \times S_K \times S_H) \\ (\downarrow\ R_{C_U} = f(R_{C_N})) \\ c_U: [P, V, M, C] \end{vmatrix}$$

Eq. 2.1.6: Light-Matrix

The matrix should be read from the top row down to the bottom row as indicated by the ↓ between rows, and suggests a series of transformations leading from the ubiquitous nature of light implicit in a point – presence, power, knowledge, harmony - to the seeming diversity of matter observed at the layer of reality U which is fundamentally the same presence, power, knowledge, and harmony projected into another form of itself.

The first transformation is summarized by Equation 2.1.7:

$$R_{C_K} = f(R_{C_\infty})$$

Eq. 2.1.7: *Light-Matrix First Transformation*

This is suggesting that the reality at the layer specified by the speed of light c_K, R_{C_K} is a function of the reality at the layer specified by the speed of light c_∞. This transformation translates the essential nature of a point into the sets described in Equation 2.1.4.

The second transformation is summarized by Equation 2.1.8:

$$R_{C_N} = f(R_{C_K})$$

Eq. 2.1.8: *Light-Matrix Second Transformation*

This is suggesting that the reality at the layer specified by the speed of light c_N, R_{C_N} is a function of the reality at the layer specified by the speed of light c_K. This transformation combines elements of the sets into unique seeds as suggested by Equation 2.1.5.

The third transformation is summarized by Equation 2.1.9:

$$R_{C_U} = f(R_{C_N})$$

Eq. 2.1.9: *Light-Matrix Third Transformation*

This is suggesting that the reality at the layer specified by the speed of light c_U, R_{c_U} is a function of the reality at the layer specified by the speed of light c_N. This transformation builds on the unique seeds suggested by (2.1.5) to create the diversity of U as specified by (2.1.1).

In this framework the notion of wave-particle duality hence may become complementary block-field-wave-particle "quadrality" where block refers to phenomenon resident to ∞, field to phenomenon resident to N, wave to phenomenon resident to K, and particle to phenomenon resident to U.

The block is all the reality always present behind the surface and is captured by (2.1.3) or the top line in the Light-Matrix as expressed in (2.1.6). The field is captured by (2.1.4) or the second line from the top in (2.1.6) and can be thought of as layers of possibility existing in each of the sets. The wave is captured by the creation of seeds represented by (2.1.5) or by the third line in (2.1.6). The particle that is apparently disconnected from the whole is captured by (2.1.1) or the bottom line in (2.1.6).

So what we arrive at is a fundamental Light-Matrix that suggests key dynamics for different layers of Light. Quanta, being at the interface of Layer U and Layer N are so positioned as being subject to multiple sets of dynamics.

Chapter 2.2: Nature of a Point-Instant in a System

The 'point-instant' captures the inherent innovation that appears to exist in the system and is represented by the top-line in Equation (2.1.6), the Light-Matrix. This innovation is a function of the properties of light derived in Equation (2.1.3) - presence, power, knowledge, and harmony – and reproduced here for convenience:

c_∞: [*Presence, Power, Knowledge, Harmony*]

The 'point' aspect of the point-instant suggests the space-dimension and that space is seeded with the possibilities inherent in the properties of presence, power, knowledge, and harmony. The 'instant' aspect suggests the time-dimension and gives insight into the process of emergence that the possibilities in space progressively surface as.

In its point-instant, presence-power-knowledge-harmony wholeness, presence allows emergences to continue to develop as per the possibilities implicit in the past-present-future or physical-vital-mental pathway.

Beginning to translate this into an equation, the notation $System_{Pr}$ is given to system-presence. This system-presence is true across any considered Time-Space continuum starting from a time-space boundary '0' to a time-space boundary 'N'. This notion is characterized by the notation $TS_{0 \to N}$. Within that boundary from 0 to 'N', the 'presence' is such that it will always seize an opportunity to cause a shift from the physical-leading to the vital-leading, and from the vital-leading to the mental-leading.

The notion that the presence seizes on opportunity as characterized by the notation:

43

Presence
↓
Opportunity

The shift from physical-leading to vital-leading and vital-leading to mental-leading is characterized by:

$$P_L \rightarrow V_L$$
$$V_L \rightarrow M_L$$

Hence in this approach it is suggested that:

$$System_{Pr} \equiv TS_{0 \rightarrow N} \begin{bmatrix} Presence \\ \downarrow \\ Opportunity \end{bmatrix} \begin{vmatrix} P_L & \rightarrow & V_L \\ V_L & \rightarrow & M_L \end{vmatrix}$$

But there is something else about this presence as well. All other developments take place in it. That is, it provides a container of sorts in which the plays of system-power, system-knowledge, and system-harmony/system-nurturing can take place. This notion is summarized by the notation:

$$Container \begin{bmatrix} System_P \\ System_K \\ System_N \end{bmatrix}$$

Hence, combining these various components, an equation for 'system-presence', Equation 2.2.1, arises:

$$System_{Pr}$$
$$\equiv TS_{0 \rightarrow N} \begin{bmatrix} Presence \\ \downarrow \\ Opportunity \end{bmatrix} \begin{vmatrix} P_L & \rightarrow & V_L \\ V_L & \rightarrow & M_L \end{vmatrix} \& Container \begin{bmatrix} System_P \\ System_K \\ System_N \end{bmatrix}$$

Eq 2.2.1: *System Presence*

In its point-instant, presence-power-knowledge-harmony wholeness, power allows emergences to continue to happen in spite of tremendous oppositions of all kinds; this too, regardless of field or area.

Constructing an equation for system-power, the notation $System_p$ is used to represent system-power. Any endeavor will always be met with resistances of various kinds. The resistances that arise along the physical dimension are referred to as P_R. The resistances that arise along the vital dimension are referred to as V_R. The resistances that arise along the mental dimension are referred to as M_R. In the fruition of any endeavor one or all of these types of resistances may arise. Further, resistance of one kind often feeds on resistance of another kind, and to generalize the resistances encountered in an endeavor may be characterized as the product of the three types of resistance:

$$P_R * V_R * M_R$$

These resistances arise across any considered Time-Space boundary from 0 to 'N', and therefore it may be said that the power of the system is such that:

$$power > \sum_{TS=0}^{N} P_R * V_R * M_R$$

An equation for 'system-power', Equation 2.2.2, hence, is the following:

$$System_P \equiv power > \sum_{TS=0}^{N} P_R * V_R * M_R$$

Eq 2.2.2: System Power

In its point-instant, presence-power-knowledge-harmony wholeness, knowledge orchestrates emergences to continue to happen by leveraging the right instruments and circumstances.

Translating this into an equation, the notation, $System_K$, is used for system-knowledge. This $System_K$ is such that it leverages the right instrumentation and circumstance to bring about the progress that is possible. This concept of 'instrumentation' is denoted by the subscript 'I'. The concept of 'circumstance' is denoted by the subscript 'C'. Both instrumentation and circumstance can be of a physical, vital, or mental type and this possibility is denoted by:

$$\begin{bmatrix} P_{I,C} \\ V_{I,C} \\ M_{I,C} \end{bmatrix}$$

Further, the notion that the 'knowledge' is such that it 'leverages' the right instrumentation and circumstance is depicted by:

$$\begin{matrix} Knowledge \\ \downarrow \\ Leverage \end{matrix}$$

This act of leveraging results in a fundamental shift so that the physical-leading yields to the vital-leading, and the vital-leading yields to the mental-leading. Hence:

$$\begin{matrix} Knowledge \\ \downarrow \\ Leverage \end{matrix} \begin{bmatrix} P_{I,C} \\ V_{I,C} \\ M_{I,C} \end{bmatrix} \rightarrow \begin{bmatrix} P_L & \rightarrow & V_L \\ V_L & \rightarrow & M_L \end{bmatrix}$$

Since this behavior may exist across any Time-Space continuum an equation for system-knowledge, Equation 2.2.3, is suggested:

$$System_K \equiv TS_{0 \rightarrow N} \begin{bmatrix} Knowledge \\ \downarrow \\ Leverage \end{bmatrix} \begin{bmatrix} P_{I,C} \\ V_{I,C} \\ M_{I,C} \end{bmatrix} \rightarrow \begin{bmatrix} P_L & \rightarrow & V_L \\ V_L & \rightarrow & M_L \end{bmatrix}$$

Eq 2.2.3: System Knowledge

In its point-instant, presence-power-knowledge-harmony wholeness, harmony or nurturance allows emergences to continue to happen with more and more degrees of freedom to come to the surface.

This characteristic of this implicit-nurturing may be referred to as 'system-nurturing'. Like the other characteristics it is suggested to exist across a Time-Space continuum. This is depicted by:

$$TS_{0 \to N}$$

There is an action of nurturing such that any state is always advanced to a higher level. This is depicted by:

$$\coprod_{Nurturing} \begin{pmatrix} P_- & M_+ \\ V_- & V_+ \\ M_- & P_+ \end{pmatrix}$$

Hence, there is a 'union', depicted by 'U' that 'nurtures' the negatives towards their positives.

Further, there is an increasing action of nurturing such that the possibility of integration is always increased to form a larger and larger basis. This increasing basis is depicted as being modulated by the polar coordinates 'r' and 'θ', where r is the radius which increases from an initial value of '0', and 'θ' is an angle from '0' to '360'.

This notion of an increasing of 'r' and 'θ' is reinforced by the relatively recent phenomena of 'Swift Trust' as a form of trust occurring in temporary organizational structures, which can include quick starting groups or teams.

Hence, the equation of system-nurturing, Equation 2.2.4, is depicted as:

$$System_N \equiv TS_{0 \to N} \left(\coprod_{Nurturing} \begin{pmatrix} P_- & M_+ \\ V_- & V_+ \\ M_- & P_+ \end{pmatrix} mod\ (r, \theta) \right)$$

Eq 2.2.4: System Nurturing

It is suggested that these four characteristics exist across

any system, and to denote this it is generalized that every point in any system is embedded with this four-fold intelligence. It is suggested that this four-fold intelligence is resident in every instant-spot of the system. It is suggested that to be able to leverage or activate this four-fold intelligence at will is the ultimate act of innovation.

Such a depiction of models, as Bar-Yam states in "From Big Data to Important Information" (Bar-Yam, 2016), 'is "valid" only because of the irrelevance of details…If we want to say anything meaningful about a system—meaningful in the sense of scientific replicability or in terms of utility of knowledge—the only description that is important is one that has universality, that is, is independent of details. There is no utility to information that is only true in a particular instance. Thus, all of scientific inquiry should be understood as an inquiry into universality—the determination of the degree to which information is general or specific.'

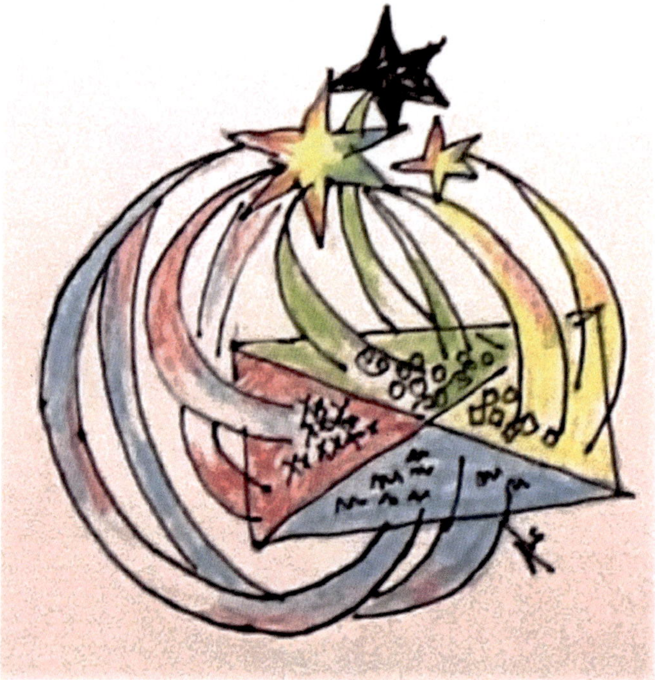

Chapter 2.3: Architectural Forces

The characteristics embedded in a point as in (2.1.3) suggest a possibility that is hard to fathom. One can only glimpse the extraordinary nature embedded in a point. And yet it can be suggested that this extraordinary nature is barely visible unless the right analytical lens of the sort being suggested here is first set up. Further, it is suggested that this extraordinary nature is responsible for a broader set of architectural forces that exist behind the visible face of things.

Hence, system-presence, system-power, system-knowledge, and system-nurturing that define the nature of every point in our system, become more tangible as a broader set of architectural forces that emanate from each of them.

Considering system-presence, here is a characteristic that appears to be everywhere (Malik, 2009) at the service of all the constructs that develop within it. There is a diligence and perseverance by which any opportunity for progress is seized. Further, if one considers the extraordinary detail that appears in any construct, whether an atom, a body, a planet, or a galaxy, one is struck by the high degree of perfection that surfaces in this presence.

So if one contemplates the nature of this system-presence there is a set of forces that surface. Depicting such a set as $S_{System_{Pr}}$, one can arrive at elements such as Service, Perfection, Diligence, Perseverance, amongst others, that are part of this set. Hence, the set can be described by Equation 2.3.1:

$$S_{System_{Pr}} \ni [Service, Perfection, Diligence, Perseverance, ...]$$

Eq 2.3.1: Set of System Presence

Similarly, considering the characteristic of system-power, one can hypothesize that there is a family of forces that emanates from it. The kinds of forces may be thought of as Power, Courage, Adventure, Justice, amongst others. The set for system-power can hence be depicted by Equation 2.3.2:

$$S_{System_P} \ni [Power, Courage, Adventure, Justice, ...]$$

Eq 2.3.2: Set of System Power

Similarly, considering the system-knowledge as the root of various powers that emanate from it, one may characterize the set for system-knowledge by Equation 2.3.3:

$$S_{System_K} \ni [Wisdom, Law Making, Spread of Knowledge ...]$$

Eq 2.3.3: Set of System Knowledge

The set for system-nurturing is depicted by Equation 2.3.4:

$$S_{System_N} \ni [Love, Compassion, Harmony, Relationship ...]$$

Eq 2.3.4: Set of System Nurturing

Chapter 2.4 describes how unique signatures or seeds for any type of organization can be built by leveraging the elements in the sets of architectural forces.

Chapter 2.4: Uniqueness of Organizations

The hypothesis is that every organization, whether an atom, cell, person, team, corporation, market, or country is unique and that this uniqueness can be specified in terms of elements of the derived sets for power, knowledge, presence, and nurturing. This hypothesis for uniqueness stems from observations at multiple levels.

At the sub-atomic level Nobel Laureate Wolfgang Pauli's 'Pauli Exclusion Principle' states that no two similar fermions, which include fundamental particles with half-integer spin such as protons, neutrons, and electrons, can occupy the same quantum states simultaneously (Pauli, 1964). Spin has to do with the angle that the particle has to rotate through before being symmetrical with its original state. Half-integer spin particles need to rotate through 720 degrees before being symmetrical with their original state. The implication of the Pauli Exclusion Principle is that fundamental structure and consequently stability comes into being at the atomic level, which as is evident in the periodic table allows the separation of function related to form. This stability related to the underlying structure of atoms implies the basis of uniqueness and diversity. In the absence of the Exclusion Principle matter would just be a dense soup (Hawking, 1988) with particles occupying overlapping space.

At the observable level uniqueness is evident from the immense diversity of distinct species on earth (Mora, 2011) estimated to be over 2 million, and further the uniqueness of every member of each species. This member-level uniqueness is suggested by the difference in non-coding regions of the DNA that may vary in their sequence by about 1 to 4 percent, which in turn result in unique protein binding sequences of each human (Snyder, 2010), as an example, which in turn results in unique observable qualities.

At the astronomical level Einstein's Special Theory of Relativity (Einstein, 1995) suggests that every coordinate system potentially has its own space-time rendering as opposed to there being one absolute space and time. This implies the notion of uniqueness as an implicit property of space.

The four properties explored in the Chapter 2.1 and 2.2 define the source of that innovation. From this source emanate 4 sets of forces that suggest the boundaries of that innovation.

Assuming then that the fount of uniqueness is system-presence, a general equation for organizations that belong to the family of system-presence can be derived. Such uniqueness can be depicted as Sig_x where the subscript 'x' refers to the source family, and 'Sig' or signature to 'uniqueness'. Hence the uniqueness of an organization in the family of system-presence would be notated by $Sig_{System_{Pr}}$.

In line with the development of properties of a point and the precipitating architectural forces as discussed in Chapter 2.1 and 2.2 respectively, an approach to constructing such uniqueness is to assume a primary factor X that drives the uniqueness that belongs to the set $S_{System_{Pr}}$. Further, assume that the uniqueness is qualified by a number of secondary factors Y that may belong to any of the 4 sets - $S_{System_{Pr}}, S_{System_P}, S_{System_K}, S_{System_N}$. The primary factor X would have a greater weightage than any of the secondary factors Y. The weightage of X hence could be depicted by the number 'a', and the weightage of Y a number 'b_{o-n}', such that a > b. Further, the secondary element can repeat from '0 – n' times, and is hence depicted as $\overline{Yb_{0-n}}$.

The equation, Equation 2.4.1, hence for a unique organization derived from the family of system-presence is:

$$Sig_P = Xa + \overline{Yb_{0-n}} \text{ where } \begin{bmatrix} X \in [S_{System_{Pr}}] \\ Y \in [S_{System_{Pr}}, S_{System_P}, S_{System_K}, S_{System_N}] \\ a, b \text{ are integers}; a > b \end{bmatrix}$$

Eq 2.4.1: System Presence Based Unique Organization

Similarly, an equation, Equation 2.4.2, for a unique organization derived from the family of system-power is:

$$Sig_V = Xa + \overline{Yb_{0-n}} \text{ where } \begin{bmatrix} X \in [S_{System_P}] \\ Y \in [S_{System_{Pr}}, S_{System_P}, S_{System_K}, S_{System_N}] \\ a, b \text{ are integers}; a > b \end{bmatrix}$$

Eq 2.4.2: System Power Based Unique Organization

An equation, Equation 2.4.3, for a unique organization derived from the family of system-knowledge is:

$$Sig_M = Xa + \overline{Yb_{0-n}} \text{ where } \begin{bmatrix} X \in [S_{System_K}] \\ Y \in [S_{System_{Pr}}, S_{System_P}, S_{System_K}, S_{System_N}] \\ a, b \text{ are integers}; a > b \end{bmatrix}$$

Eq 2.4.3: System Knowledge Based Unique Organization

An equation, Equation 2.4.4, for a unique organization derived from the family of system-nurturing is:

$$Sig_I = Xa + \overline{Yb_{0-n}} \text{ where } \begin{bmatrix} X \in [S_{System_N}] \\ Y \in [S_{System_{Pr}}, S_{System_P}, S_{System_K}, S_{System_N}] \\ a, b \text{ are integers}; a > b \end{bmatrix}$$

Eq 2.4.4: System Nurturing Based Unique Organization

The four preceding equations can be generalized by Equation 2.4.5:

$$Sig = Xa + \overline{Yb_{0-n}} \text{ where } \begin{bmatrix} X \in [S_{System_{Pr}}, S_{System_P}, S_{System_K}, S_{System_N}] \\ Y \in [S_{System_{Pr}}, S_{System_P}, S_{System_K}, S_{System_N}] \\ a, b \text{ are integers}; a > b \end{bmatrix}$$

Eq 2.4.5: Generalized Equation for Unique Organization

Having considered the structure of uniqueness, the next question is how does such uniqueness emerge? This is discussed in Chapter 2.5.

Chapter 2.5: Emergence of Uniqueness

While the uniqueness of organizations as represented by the Signature is a seed, like any seed there is a process for its emergence (Kaufmann, 1995; Portugali, 2012; Yates, 2012), and the uniqueness will often be hidden or very much behind the scene until certain conditions are fulfilled (Malik, 2009).

The implicit nature of Time and Space suggest a universal developmental model that provides a cue as to the process for emergence (Deep Order Mathematics Videos, 2016). In this model the four sets of architectural forces already described form a pool in space, as it were, from which possibility arises. Possibility itself is unique from point to point and is governed by the Equation for Uniqueness (Equations 2.4.1 through 2.4.5) described in the previous chapter.

Hence it is observed that initially the uniqueness takes a 'physical' form, moving on to a 'vital' form, and then onto a 'mental' form. Once the orientations implicit in each of these phases are assimilated, then the uniqueness takes on an 'integral' form. The integral form is a threshold phase, and allows the uniqueness suggested by the Signature to emerge in fuller force or in its 'force' form.

The final phase is the 'contextual form' that allows the signature to act with impunity within a considered context.

Mathematically, if an organization exists at the physical phase, it may be suggested that its signature or uniqueness is modulated by the constant 'π'. π is the seed of a circle or sphere and can be thought of as defining behavior that is tightly bound. Within such a tightly bound volume it will likely not even be apparent what the uniqueness of an organization necessarily is. Assuming the uniqueness to be defined by the derived question *Sig*, the physical-level (P) behavior can be described by the following equation-segment where 'mod' signifies modulated-by:

$P: Sig * mod\,(\pi)$

If an organization exists at the vital level, it may be suggested that its uniqueness is modulated by the Euler-constant 'e'. e is at the root of exponential behavior. The vital by definition is about assertive and aggressive growth the symbol of which is 'e'. Hence vital-level (V) modulation (represented by 'mod') can be described by the following equation-segment:

$V: Sig * mod\,(e)$

If an organization exists at the mental level, it may be suggested that its uniqueness is modulated by the Gaussian Distribution 'G'. G summarizes rational behavior with a key direction followed by most, and directions more on the edge followed by outliers. Mental-level dynamics are arguably quite similar, and it can be suggested are best modeled by such a distribution (Salkind, 2007). Mental-level (M) modulation (mod) can hence be described by the following equation-segment:

$M: Sig * mod\,(G)$

The physical, the vital, and the mental levels are orientations in which patterns of perceiving, being, behaving are set in their ways. Each pattern has its purpose and its limitation and it can be argued that being able to learn from each orientation and yet being able to move beyond that, is the next logical step in any developmental model. The integral level hence, is about being able to leverage each of the patterns that naturally arise at the three preceding levels at will, and about further, being able to integrate these and arrive at new ways of perceiving and being.

Mathematically such behavior may be represented as being an integrative function ($\int x$) where 'x' is the ability to move between the patterns emanating from G, e, π, at will, represented by $\overline{G, e, \pi}$. Integral-level (I) modulation (mod) of uniqueness (Sig) can hence be represented by the following equation-segment:

$$I: Sig * mod \left(\int \overline{G, e, \pi} \right)$$

The condition of overcoming any fixed and limiting patterns is the prerequisite for the emergence of 'Force' or for entering into the force-level. At this level the uniqueness behind the particular development being considered can emerge in its purity and become a truly creative dynamic. This aspect of creativity that is in a sense not bound by circumstance may be represented by the constant 'c', the speed of light in a vacuum (Perkowitz, 2011) which is an upper limit of the layer that systems practically operate in. Force-level (F) modulation (mod) of uniqueness (Sig) can hence be represented by the following equation-segment:

$$F: Sig * mod (c)$$

Once the signature of an organization arises and continues to exercise itself in its purity, it achieves contextual-mastery (C) and is able to exercise itself as though the context it is acting in, that can vary in scale and complexity, were all of the same substance as itself. This equality may be represented by the integrative function '$\int = 1$'. The equation-segment that notates this contextual-level (C) modulation (mod) applied to organizational uniqueness (Sig) is hence:

$$C: Sig * mod \left(\int = 1\right)$$

Piecing all the equation-segments together the equation for the emergence of uniqueness (Sig_E), where 'X' can be any of the discussed modulations at the respective development-model levels (P, V, M, I, F, C), is hence summarized by Equation 2.5.1:

$$Sig_E = X \begin{vmatrix} C: Sig * mod \left(\int = 1\right) \\ F: Sig\ mod\ (c) \\ I: Sig\ mod\ \left(\int \overline{G, e, \pi}\right) \\ M: Sig * mod\ (G) \\ V: Sig * mod\ (e) \\ P: Sig * mod\ (\pi) \end{vmatrix}$$

Eq 2.5.1: *Emergence of Uniqueness*

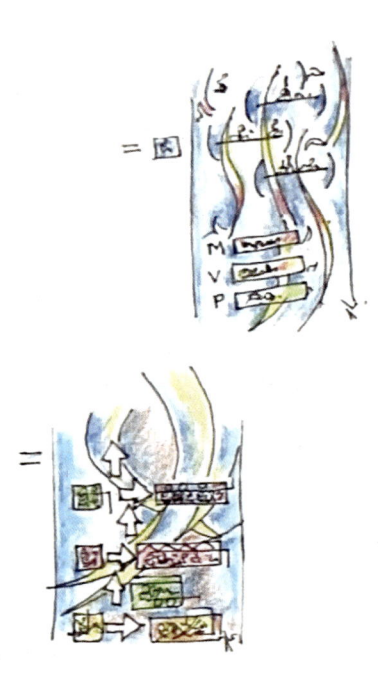

The power of virtual worlds in engaging people and promoting learning as described in "Why Virtual World Matter" (Thomas & Brown, 2009) can be seen as parallel to the emergence of uniqueness as described by Equation 2.5.1. As described by the authors there is a process of selectively leaving behind part of oneself (movement through the P, V, M levels) to recreate one's identity (movement through level I), and then engaging with others in a shared discourse

62

and culture (movement through levels F, C), which becomes very meaningful for people.

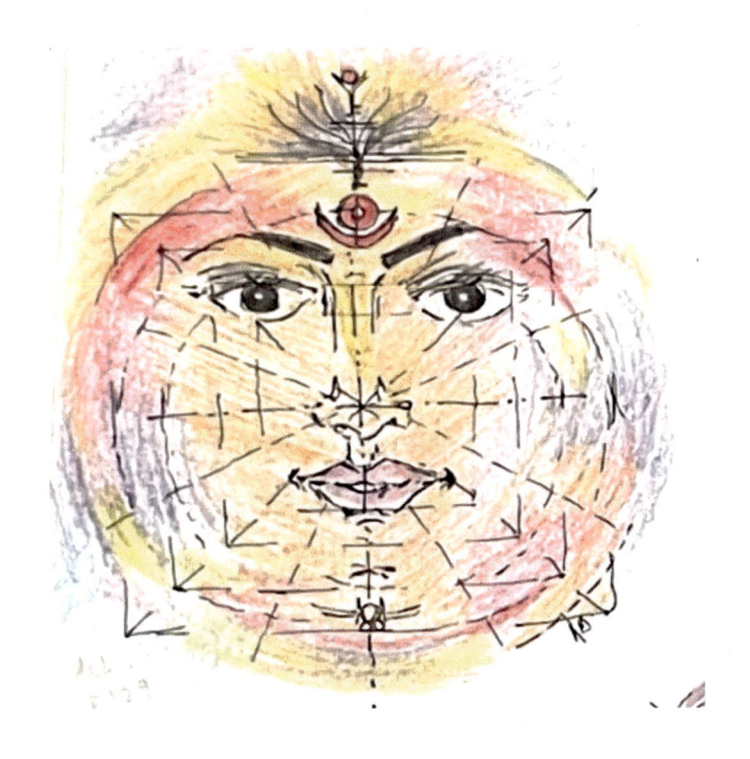

Chapter 2.6: Inherent Dynamics of Any System

So far the inherent innovation that exists at the system level and summarized by the nature of a point has been considered. Further, how this deep fount of innovation is present everywhere and how sets that make more practical the range of creative forces available in each of the four components of a point have also been considered. These architectural forces further define the possibility inherent in any system. Leveraging these sets of forces ,an equation for the uniqueness of an organization regardless of scale, was arrived at.

In some sense the precipitation of innovation from the barely perceptible nature of the ubiquitous point, to how this reveals a play of forces, to how organizations take their seed and grow from these forces, has been traced.

It is useful to now turn full-circle to return to the initial orientations that allowed so much to be suggested about the nature of innovation in the first place. It is useful to look deeper into the nature of the physical, the vital, the mental, and the integral, and to derive equations that in effect will provide further insight into the dynamics of innovation inherent in these orientations.

Hence, starting with the physical, an equation, Equation 2.6.1, is summarized as:

$$Physical = \begin{bmatrix} M_3 \rightarrow System_{Pr} \\ (\uparrow F \rightarrow I) \\ M_2 \rightarrow S_{Systemi_{Pr}} \\ (\uparrow Sig \rightarrow F) \\ M_1 \rightarrow Sig_P \\ (\uparrow > P_P) \\ U \rightarrow Physical_U \end{bmatrix} TC \rightarrow Physical_T$$

$$\text{Where} \begin{bmatrix} Physical_U \ni [inertia, lethargy, status\ quo, ...] \\ Physical_T \ni [adaptability, durability, strength, ...] \end{bmatrix}$$

Eq 2.6.1: Inherent Dynamism in Physical

Essentially this equation is laying out the conditions of moving from the untransformed or negative physical state represented by $Physical_U$ to the transformed or positive physical state represented by $Physical_T$.

The first matrix should be read from the bottom to the top:

$$\begin{bmatrix} M_3 \rightarrow System_{Pr} \\ (\uparrow F \rightarrow I) \\ M_2 \rightarrow S_{System_{Pr}} \\ (\uparrow Sig \rightarrow F) \\ M_1 \rightarrow Sig_P \\ (\uparrow > P_P) \\ U \rightarrow Physical_U \end{bmatrix}$$

Hence, at the bottom is the starting point '$U \rightarrow Physical_U$' which identifies the default or untransformed (U) level of the physical. The next row up, $(\uparrow > P_P)$, states that when the patterns of the untransformed physical (P_P) have been overcome (>), movement to the next level ($\uparrow$) is facilitated. Breaking through to the next level, $M_1 \rightarrow Sig_P$, allows its dynamics to become active. Hence, the signature or uniqueness of the physical (Sig_P) becomes active at meta-level 1 (M_1). As this signature becomes more like a Force ($Sig \rightarrow F$), the conditions for breakthrough ($\uparrow$) to the next level are achieved. This next level is referred to as meta-level 2 (M_2), and indicates that the architectural forces represented by the set of system-presence ($S_{System_{Pr}}$) have become more consciously active. When this Force becomes Integral ($F \rightarrow I$) then the conditions for breakthrough ($\uparrow$) to the next level are achieved. The next level is notated as M_3 for meta-level 3, and the dynamics here indicate that the equation for system-presence becomes active. Becoming active basically means that the respective meta-level dynamic begins to act at the once 'untransformed' level (U) further modifying it. Modification or transformation began when M_1 became active. Transformation is accelerated when M_2 becomes active, and even further accelerated when M_3 becomes active. Note that the matrix is essentially semiotic in that adjacent levels exist in paired relationships and inter-related functionality (Usó-Doménech et al, 2016).

The notion of meta-layers is being explored by contemporary physicists and Erwin Laszlo in his book Self-Actualizing Cosmos (Laszlo, 2014) summarizes some of these developments: "Physicists describe the domain that underlies and embeds the particles, fields, and forces of the universe variously as quantum vacuum, physical spacetime, nuether, zero-point field, grand-unified field, cosmic plenum, or string-net liquid." Note that 'nuether' refers to a sub-quantum level of reality (Pearson, 1997). Laszlo goes on to describe a revolutionary discovery by Nika Arkani-Hamed of Princeton's Institute of Advanced Study, of a geometrical object, the amplituhedron (Arkani-Hamid et al, 2012), which is not in space-time but governs space-time so that it "appears that spatio-temporal phenomena are the consequence of geometrical relationships in a deeper dimension of physical reality". A deeper dimension of a physical layer suggests synonymity with a meta-layer.

The rate of the transformation can be better envisioned when considering action of the Transformation Circle, or TC. The TC can be thought of as 4 concentric circles, with M_3 at the center. M_3 is surrounded by M_2, which is surrounded by M_1. The outer circle is U. If TC is considered to be a clock, than at time 't = 0', the physical' can be thought of as being entirely in U. The clock starts ticking only when some initial patterns P_P are overcome

($>P_P$). From this point on as time proceeds the conditions for breakthrough become riper, and a sinusoidal wave begins to integrate more of the concentric circles together. The sinusoid wave (sin) is itself modulated by an euler function, e^x, where 'x' is determined by the strength to overcome patterns (↑) which will likely vary over time but will likely tend to be positive once the clock has started ticking because of the joy experienced with progressive movement. Being that the limit is the outer boundary of the concentric circles, there is further modulation by π until the 4 concentric circles have been integrated. TC, hence, may be represented by Equation 2.6.2:

$$TC \equiv (> P_P) \to \text{mod}(\sin, e^x, \pi)$$

Eq 2.6.2: Transformation Circle

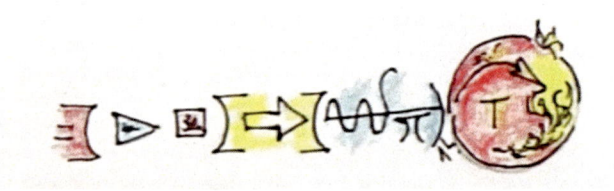

Hence, the initial nature of the physical that may be characterized by the set comprising of elements such as, lethargy, acceptance of the status quo, amongst other such elements ($Physical_U \ni [inertia, lethargy, status\ quo, ...]$), transforms into a physical more characterized by elements such as adaptability, durability, strength, and so on ($Physical_T \ni [adaptability, durability, strength, ...]$). This transformation represents the inherent innovation-dynamic within the Physical.

Similarly, the equation for the 'Vital', Equation 2.6.3, also shows the built-in transformation that represents the innovation-dynamic within the vital:

$$Vital = \begin{bmatrix} M_3 \to System_P \\ (\uparrow F \to I) \\ M_2 \to S_{System_P} \\ (\uparrow Sig \to F) \\ M_1 \to Sig_V \\ (\uparrow > P_V) \\ U \to Vital_U \end{bmatrix} TC \to Vital_T,$$

$$\text{Where} \begin{bmatrix} Vital_U \ni [aggression, self\ centeredness, exploitation, ...] \\ Vital_T \ni [energy, support, adventure, enthusiasm, ...] \end{bmatrix}$$

Eq 2.6.3: *Inherent Dynamism of Vital*

The equation for the 'Mental', Equation 2.6.4, is similarly summarized as:

$$Mental = \begin{bmatrix} M_3 \rightarrow System_S \\ (\uparrow F \rightarrow I) \\ M_2 \rightarrow S_{Systems} \\ (\uparrow Sig \rightarrow F) \\ M_1 \rightarrow Sig_M \\ (\uparrow > P_M) \\ U \rightarrow Mental_U \end{bmatrix} TC \rightarrow Mental_T$$

Where $\begin{bmatrix} Mental_U \ni [fixation, fundamentalism, fragmentation, ...] \\ Mental_T \ni [understanding, imagination, inspiration, ...] \end{bmatrix}$

Eq 2.6.4: *Inherent Dynamism of Mental*

The equation for the 'Integral', Equation 2.6.5, is similarly summarized as:

$$Integral = \begin{bmatrix} M_3 \to System_N \\ (\uparrow F \to I) \\ M_2 \to S_{System_N} \\ (\uparrow Sig \to F) \\ M_1 \to Sig_I \\ (\uparrow > P_{I)} \\ U \to Integral_U \end{bmatrix} \quad TC \to Integral_T$$

Where $\begin{bmatrix} Integral_U \ni [possession, usurpation, hidden\ agendas, ...] \\ Integral_T \ni [appreciation, shift\ POV, MPV, synthesis, ...] \end{bmatrix}$

Eq 2.6.5: *Inherent Dynamism of Integral*

The preceding equations can be generalized by Equation 2.6.6:

$$Innovation_{orientation-x}$$
$$= \begin{bmatrix} M_3 \to System_X \\ (\uparrow F \to I) \\ M_2 \to S_{System_X} \\ (\uparrow Sig \to F) \\ M_1 \to Sig_x \\ (\uparrow > P_{x)} \\ U \to x_U \end{bmatrix} \quad TC \to x_T, where \begin{bmatrix} x_U \ni [...] \\ x_T \ni [...] \end{bmatrix}$$

Eq 2.6.6: *Generalized Equation of Innovation*

In this generalized equation, $Innovation_{orientation-x}$, refers to the inherent innovation within a specific orientation. Orientation refers to the physical, the vital, the mental, or the integral.

Further, the notion of a core-matrix can be summarized by the following equation, Equation 2.6.7:

$$Core\_matrix = \begin{bmatrix} M_3 \rightarrow System_X \\ (\uparrow F \rightarrow I) \\ M_2 \rightarrow S_{System_X} \\ (\uparrow Sig \rightarrow F) \\ M_1 \rightarrow Sig_x \\ (\uparrow > P_{x)} \\ U \rightarrow x_U \end{bmatrix}$$

Eq 2.6.7: Core Matrix

One of the corollaries of the Generalized Equation of Innovation is that if the source of innovation is more influenced by a meta-level there will be simultaneity of innovation or emergence that becomes apparent at U. The higher the meta-level, the more likely that this simultaneity will be wider spread. In his book Where Good Ideas Come From: The Natural History of Innovation (Johnson, 2010) Johnson states: "A brilliant idea occurs to a scientist or inventor somewhere in the world, and he goes public with his remarkable finding, only to discover that three other minds had independently come up with the same idea in the past year." He refers to an essay "Are Inventions Inevitable" (Ogburn & Thomas, 1922) which uncovered 148 instances of similar yet independent innovation, most of them occurring within the same decade. Some examples include sunspots that were simultaneously discovered in 1611 by four scientists in four different countries, and the law of conservation of energy that was formulated separately four times in the late 1840s, amongst numerous other example.

Chapter 2.7: Qualified Determinism

Leveraging off the previous chapter a mathematical notion of qualified determinism is explored here.

A new function, Dynamic Interaction (DI) that has a

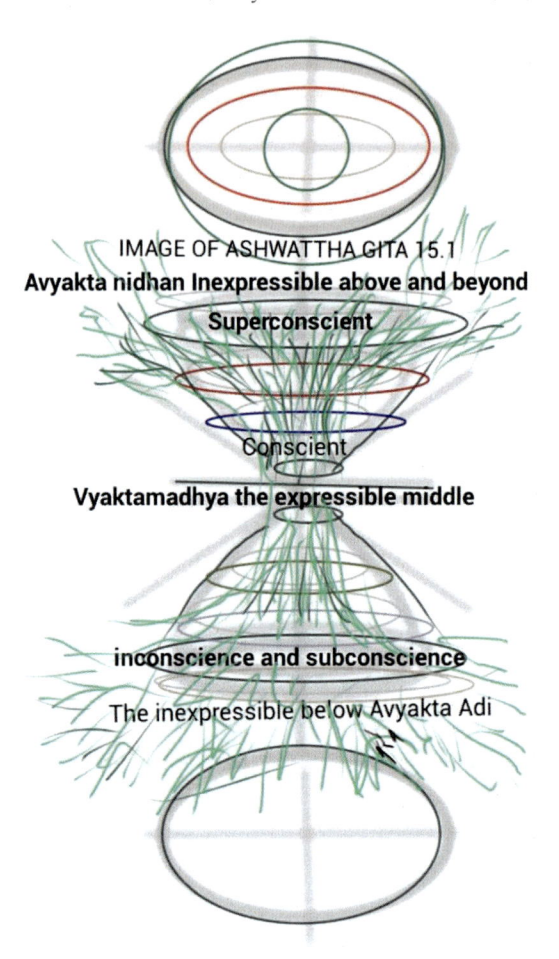

IMAGE OF ASHWATTHA GITA 15.1
Avyakta nidhan Inexpressible above and beyond
Superconscient
Conscient
Vyaktamadhya the expressible middle
inconscience and subconscience
The inexpressible below Avyakta Adi

'vertical' and a 'horizontal' component is introduced here. The vertical component is designated as DI_V and the horizontal component as DI_H. Several equations to capture the inherent dynamism that exists in each

72

orientation or state have already been derived in Chapter 2.6. These included equations for the dynamism in the physical, the vital, the mental, and the integral. The derived equations propose a model to give insight into how innovation occurs by changing the fundamental states that an organization is subject to. Several scientists, such as Prigogine (Prigogine, 1977) and others, are proposing that a system can bifurcate in unpredictable ways to create an emergent property that cannot be predicted. DI is going to propose that in fact there is a 'qualified determinism' as opposed to randomness that occurs.

This qualified determinism is the result of the relative strengths of the levels within the core-matrix identified in the derived equations and summarized by Equation 2.6.7. Hence, the generalized core-matrix, already is the following:

$$\begin{bmatrix} M_3 \rightarrow System_X \\ (\uparrow F \rightarrow I) \\ M_2 \rightarrow S_{System_X} \\ (\uparrow Sig \rightarrow F) \\ M_1 \rightarrow Sig_x \\ (\uparrow > P_{x)} \\ U \rightarrow x_U \end{bmatrix}$$

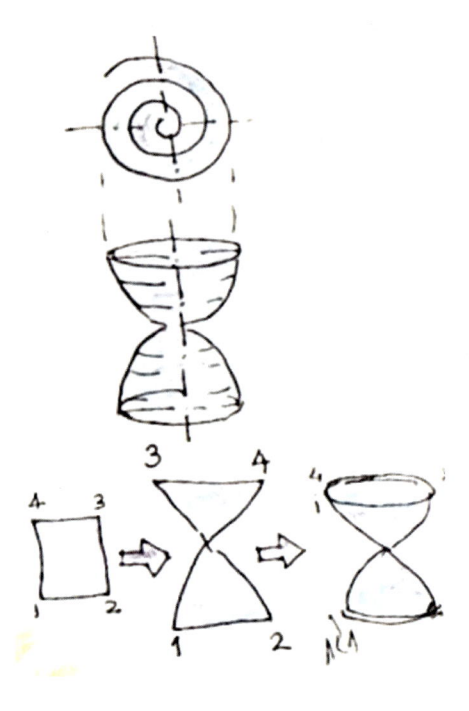

The application of the vertical component of the new function being proposed, DI_V, to this core matrix will yield the nature or 'strength' of the state (x) or orientation under consideration. If the untransformed or U layer is strongest, implying that the habitual patterns that keep an organization locked into its untransformed way of operation are still very active, then the nature of the output of DI_V, notated by x-state, will be x_U. If the habitual patterns have been overcome then the strength of the x-state increases since it is the dynamics of M_1 or Sig_x that are now active. In this case the x-state will be Sig_x. If the unique 'signature' has become a 'force', then the conditions for activation of M_2 have been put in place and the x-state will be even higher, S_{System_X}. The architectural forces active in M_2 are by definition more powerful than Sig_x that is a derivation of a set of such architectural forces. If the 'force' so acting becomes impersonal so that an organizational ego-state is

74

overcome, then the x-state will have the most strength and is characterized by $System_X$ active at M_3. Hence, DI_V applied to a core-matrix will yield the 'strength' in terms of the x-dynamic that is active. This is illustrated by the following equation, Equation 2.7.1 which can be considered to be a deductive proof in the context of this model:

$$DI_V \begin{bmatrix} M_3 \rightarrow System_X \\ (\uparrow F \rightarrow I) \\ M_2 \rightarrow S_{System_X} \\ (\uparrow Sig \rightarrow F) \\ M_1 \rightarrow Sig_x \\ (\uparrow > P_{x)} \\ U \rightarrow x_U \end{bmatrix} =>$$

x-state $\in (x_U, Sig_x, S_{System_X}, System_X)$

$Where: Strength(System_X) > Strength(S_{System_X})$
$> Strength(Sig_x) > Strength(x_U)$

Eq 2.7.1: Illustrating Action of Dynamic Interaction – vertical component

What is to be noted here is that while the action of DI_V yields a relative strength and therefore a 'single' value for the core- or x-matrix under consideration yet each x-matrix in itself could have an infinite number of possibilities. This should be clear in looking at how x_U, Sig_x, S_{System_X}, and $System_X$, were initially defined.

Hence, taking the example where x = physical:

$Physical_U \ni [inertia, lethargy, status\ quo, ...]$

As can be seen $Physical_U$, defined in Chapter 2.6, is already an infinite set with qualities similar to the ones already specified.

Similarly, Sig_P, defined in Chapter 2.4, also has an infinite variation:

$$Sig_P = Xa + \overline{Yb_{0-n}} \quad \text{where} \quad \begin{bmatrix} X \in [S_{System_{Pr}}] \\ Y \in [S_{System_{Pr}}, S_{System_P}, S_{System_K}, S_{System_N}] \\ a, b \text{ are integers}; a > b \end{bmatrix}$$

$S_{System_{Pr}}$, defined in Chapter 2.3, is also an infinite set with forces of the nature specified in the following equation:

$$S_{System_{Pr}} \ni [Service, Perfection, Diligence, Perseverance, ...]$$

And recall that in Chapter 2.2, $System_{Pr}$ has been defined as:

$$System_{Pr} \equiv TS_{0 \to N} \begin{bmatrix} Presence \\ \downarrow \\ Opportunity \end{bmatrix} \begin{bmatrix} P_L & \to & V_L \\ V_L & \to & M_L \end{bmatrix} \& \; Container \begin{bmatrix} System_P \\ System_K \\ System_N \end{bmatrix}$$

So in essence DI_V is really giving us a summary assessment of the 'level' of the x-matrix under consideration with all its infinite potentiality. An example will follow shortly.

The other component of DI, as suggested earlier in this chapter, is the horizontal component, DI_H. Just as DI_V yields a summary assessment of the level that an x-matrix is operating at, similarly DI_H yields a summary assessment of the direction that a system or organization under consideration is going to continue its development in considering the physical, the vital, the mental, and the integral orientations to be the choices.

Assuming that any organization or system is inherently unique, as this mathematical model proposes, and assuming that the infinite sets of x_U and S_{System_X} applied across the physical, vital, mental, and integral orientations respectively will account for any state that an organization can experience, then at a certain point in time any organization under consideration is going to have a direction-bias in one of the possible physical, vital, mental, or integral directions. Hence, DI_H will yield the summary direction that is going to lead an organization into its future given the current states active in it.

This summary direction is going to be yielded by considering the relative strengths of the separate core x-matrices – the physical, the vital, the mental, the integral - under consideration. The assumption is that there will be one core-matrix that will be stronger than the others.

Hence, as an example, first applying DI_V across all four x-matrices may, for example, yield the following results, with the strongest level within each x-matrix highlighted and bolded:

$$\begin{bmatrix} System_{Pr} \\ \mathbf{S_{System_{Pr}}} \\ Sig_P \\ Physical_U \end{bmatrix} \begin{bmatrix} System_P \\ S_{System_P} \\ Sig_V \\ \mathbf{Vital_U} \end{bmatrix} \begin{bmatrix} System_K \\ S_{System_K} \\ Sig_M \\ \mathbf{Mental_U} \end{bmatrix} \begin{bmatrix} System_N \\ S_{System_N} \\ \mathbf{Sig_I} \\ Integral_U \end{bmatrix}$$

Since by definition the strength of $System_x$ is greater than S_{System_x}, which is greater than Sig_x, which is greater than x_U, applying DI_H, as in Equation 2.7.2, across these x-matrices, as in the example following it will then yield the strongest direction, which in this example is the Physical:

$$DI_H \left(\begin{bmatrix} System_{Pr} \\ S_{System_{Pr}} \\ Sig_P \\ Physical_U \end{bmatrix} \begin{bmatrix} System_P \\ S_{System_P} \\ Sig_V \\ Vital_U \end{bmatrix} \begin{bmatrix} System_K \\ S_{System_K} \\ Sig_M \\ Mental_U \end{bmatrix} \begin{bmatrix} System_N \\ S_{System_N} \\ Sig_I \\ Integral_U \end{bmatrix} \right)$$
$$= Orientation_{Strongest}$$

Eq 2.7.2: Illustrating Action of Dynamic Interaction - horizontal component

Example:

$$DI_H \left(\begin{bmatrix} System_{Pr} \\ \mathbf{S_{System_{Pr}}} \\ Sig_P \\ Physical_U \end{bmatrix} \begin{bmatrix} System_P \\ S_{System_P} \\ Sig_V \\ \mathbf{Vital_U} \end{bmatrix} \begin{bmatrix} System_K \\ S_{System_K} \\ Sig_M \\ \mathbf{Mental_U} \end{bmatrix} \begin{bmatrix} System_N \\ S_{System_N} \\ \mathbf{Sig_I} \\ Integral_U \end{bmatrix} \right)$$
$$= Physical$$

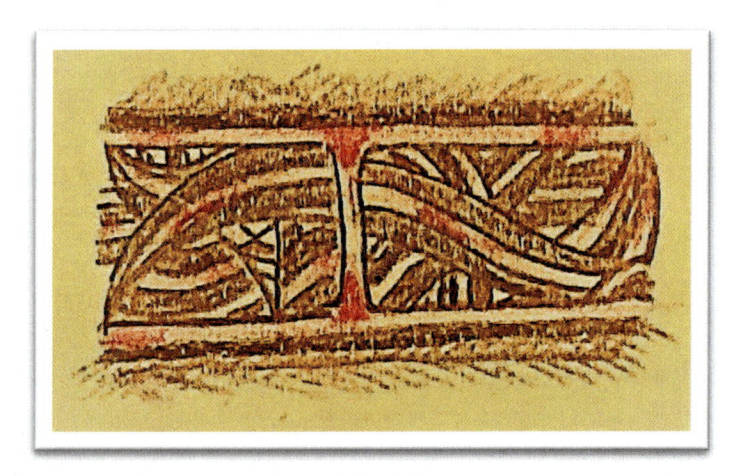

Hence, DI function will yield the following organizational direction, as in Equation 2.7.3, where 'x_matrix' is used interchangeably with 'orientation':

78

$$Org\_Dir = DI \left(\begin{bmatrix} M_3 \to System_{Pr} \\ (\uparrow F \to I) \\ M_2 \to S_{System_{Pr}} \\ (\uparrow Sig \to F) \\ M_1 \to Sig_P \\ (\uparrow > P_{P)} \\ U \to Physical_U \end{bmatrix} \begin{bmatrix} M_3 \to System_P \\ (\uparrow F \to I) \\ M_2 \to S_{System_P} \\ (\uparrow Sig \to F) \\ M_1 \to Sig_V \\ (\uparrow > P_{V)} \\ U \to Vital_U \end{bmatrix} \right.$$
$$\left. \begin{bmatrix} M_3 \to System_S \\ (\uparrow F \to I) \\ M_2 \to S_{System_S} \\ (\uparrow Sig \to F) \\ M_1 \to Sig_M \\ (\uparrow > P_{M)} \\ U \to Mental_U \end{bmatrix} \begin{bmatrix} M_3 \to System_N \\ (\uparrow F \to I) \\ M_2 \to S_{System_N} \\ (\uparrow Sig \to F) \\ M_1 \to Sig_I \\ (\uparrow > P_{I)} \\ U \to Integral_U \end{bmatrix} \right) \to$$

$x\_matrix_{strongest}$ @ $level_{strongest}$

Eq 2.7.3: Organizational Direction

Generalizing, as in Equation 2.7.4, where Org_Dir is organizational direction:

$$Org\_Dir = DI \left(\begin{bmatrix} M_3 \to System_X \\ (\uparrow F \to I) \\ M_2 \to S_{System_X} \\ (\uparrow Sig \to F) \\ M_1 \to Sig_X \\ (\uparrow > P_{P)} \\ U \to x_U \end{bmatrix}^{x=p,v,m,i} \right) \to$$

$x\_matrix_{strongest}$ @ $level_{strongest}$

Eq 2.7.4: Generalized Equation for Organizational Direction

Hence, this mathematical model is suggesting that any situation, rather than having a random outcome, has a 'qualified deterministic' outcome. In the introduction to his book "Where is Science Going?" (Planck, 1933), James Murphy points out that the reason Planck spent so much of his time giving lectures on causation was because of the trend of physicists at the time, which has continued to

the modern day, to overthrowing the principle of causation following the development of quantum theory, which he felt was misplaced. "Planck would claim", he wrote, "and so would Einstein, that it is not the principle of causation itself which has broken down in modern physics, but rather the traditional formulation of it." Murphy also quotes James Jeans (Jeans, 1932) to suggest the issue associated with causation and determinism:

"Einstein showed in 1917 that the theory founded by Planck appeared, at first sight at least, to entail consequences far more revolutionary than mere discontinuity", and here he is referring to the finding that radiant energy is not emitted in a continuous flow, but in integral quantities, or quanta, which can be expressed in integral numbers. Continuing: "It appeared to dethrone the law of causation from the position it had therefore held as guiding the course of the natural world. The old science had confidently proclaimed that nature could follow only one road, the road which was mapped out from the beginning of time to its end by the continuous chain of cause and effect; state A was inevitably succeeded by state B. So far the new science has only been able to say that state A may be followed by state B or C or D or by innumerable other states. It can, it is true, say that B is more likely than C, C than D, and so on; it can even specify the relative probabilities of B, C, and D. But, just because it has to speak in terms of probabilities, it cannot speak with certainty which state will follow which; this is a matter which lies on the knees of the gods – whatever gods there may be."

While under the apparent dynamics at the quantum level there may appear to be randomness and a dethroning of the principle of causation, the notion of a multiplicity of levels, each having its impact on the strength of an orientation and further on the consequent direction from a multiplicity of possible orientations, is being suggested here as determining the direction of any CAS, while still allowing infinite variation in the details that may define its. Hence, the positions of Planck and Einstein are vindicated when considering Equation 11.4.

Further, assuming any CAS where multiple elements are active, connected, interdependent, and emergent, it may be possible to understand, through application of calculus, as to which level is the source for change.

Hence, where N may be source of change, the rate of change of N will resolve into one of P_U, V_U, M_U, I_U, P_T, V_T, M_T, or I_T. This may be summarized by Equation 2.7.5, where y is either U or T:

$$\frac{dN}{dt} \rightarrow \begin{bmatrix} P_U & P_T \\ V_U & V_T \\ M_U & M_T \\ I_U & I_T \end{bmatrix} \rightarrow x_y, \text{where } y \in (U,T)$$

Eq 2.7.5: Establishing the Nature of the Change

If T, implying that the action of one of the meta-levels has caused transformation, then application of one of the following integrals will determine which level is the likely source for change.

Hence, for M_1, if the integral of $\frac{\partial(x_U \rightarrow x_T)}{\partial t}$ across a limited area 'a' in the vicinity of the change, is greater than some threshold value $Threshold_{Signature}$, then the signature dynamics are likely the source of change. This is summarized by Equation 2.7.6:

$$\int_0^a \frac{\partial(x_U \rightarrow x_T)}{\partial t} > Threshold_{Signature}$$

Eq 2.7.6: Signature Dynamics as the Source of Change

For M_2, if the integral of $\frac{\partial(x_U \rightarrow x_T)}{\partial t}$ across a larger area 'b' extending beyond the vicinity of the change, is greater than some threshold value $Threshold_{Architectural\ Forces}$, then the architectural forces are likely the source of change. This is summarized by Equation 2.7.7:

$$\int_0^b \frac{\partial(x_U \rightarrow x_T)}{\partial t} > Threshold_{ArchitecturalForces}$$

Eq 2.7.7: Architectural Forces as the Source of Change

For $M_{3.,}$ if the double integral of $\frac{\partial(x_U \to x_T)}{\partial t}$ across the CAS specified by 'A', and across some time 't', is greater than some threshold value $Threshold_{System\,Property}$, then the system properties are likely the source of change. This is summarized by Equation 2.7.8:

$$\int_0^t \int_0^A \frac{\partial(x_U \to x_T)}{\partial t} > Threshold_{SystemProperty}$$

Eq 2.7.8: System Properties as the Source of Change

Chapter 2.8: Equation for Light-Space-Time Emergence

Equation 2.6.6, the generalized equation for innovation, can be restated as an evolving form true for all time, as in Equation 2.8.1:

$$Innovation_{orientation-x} = \left(\begin{bmatrix} M_3 \to System_X \\ (\uparrow F \to I) \\ M_2 \to S_{System_X} \\ (\uparrow Sig \to F) \\ M_1 \to Sig_x \\ (\uparrow > P_x) \\ U \to x_U \end{bmatrix} TC \to x_T, where \begin{bmatrix} x_U \ni [\ldots] \\ x_T \ni [\ldots] \end{bmatrix} \right)_{\langle x_U | x_T \rangle}$$

Eq 2.8.1: Evolving Form of Generalized Equation of Innovation

84

But further, given that quanta are proposed to be a doorway to deeper worlds of Light, that in fact allow aspects of those worlds or layers to become active at the surface layer U, the question is when have those aspects become active in manifest time. The following suggests when:

- At time, t = 0 seconds, only M_3 is active. Recall that M_3 represents the four-fold reality present in every point-instant.
- At time, $0 \geq t > \infty$, M_2 the set of architectural forces continually gets added to, thereby increasing the size of the sets of forces.
- At time, $0 > t \geq 10^{-36}$ seconds, the equation of Innovation, $Innovation_{orientation-x}$, is such that M_1 also becomes active, while U and TC are suppressed while present. The activation of M_1 begins to result in unique expressions or signatures of the set of architectural forces, and in this case in the reality of the essentially ubiquitous EM Spectrum as a vehicle of the four-fold order that expressed itself in all that existed and in all that unfolded from that point in time on.
- At time, $t \sim 10^{-10}$ seconds, fundamental particles emerge as an essential material basis of the four architectural forces that frame all further development. As in the case of the EM Spectrum this implies the activity of M_1.
- At time, $t \sim 3 \times 10^5$ years light atoms emerge, and at time $t \sim 10^9$ years heavier atoms in the stars emerge. These also imply the continued activity of M_1.
- At time, $t \sim 13.8 \times 10^9$ years, the fourth clear expression of the same four-fold order as the

85

bases of an even more complex organization, that of cellular life and all that is founded on it becomes clear. This time-point will be represented by the notation $t \sim E_{Cell}$, where 'E' stands for emergence. This too implies the activity of M_1. But further, TC becomes active here, with an implicit direction of operation from M_3 to U. Note that the sets of architectural forces specified by M_2 continue to increase the number of elements they comprise of as the complex interaction between the layers continues.

- At time $t > 13.8 \times 10^9$ years, human-beings, and more complex social organizations emerge. Here TC acts with an implicit direction of operation from U to M_3. This time-point will be represented by $t \sim E_{Human}$.

Based on the aforementioned description Equation 2.8.2 for Emergence true of any space-time scale may be generalized as the following:

$$Emergence_{space-time}$$

$$= \begin{vmatrix} \begin{bmatrix} M_3 \to System_X \\ (\uparrow F \to I) \\ M_2 \to S_{System_X} \\ (\uparrow Sig \to F) \\ M_1 \to Sig_x \\ (\uparrow > P_x) \\ U \to x_U \end{bmatrix}_{Space} \\ \begin{bmatrix} M_3 : All\ t \\ \downarrow \\ M_2 : 0 \geq t > \infty \\ \downarrow \\ M_1 : 0 > t > \infty \\ \downarrow \\ U \to \begin{matrix} t \sim E_{Cell}; TC: M_3 \to U \\ t \sim E_{Human}; TC: U \to M_3 \end{matrix} \\ TC \to x_T, where\ \begin{bmatrix} x_U \ni [\ldots] \\ x_T \ni [\ldots] \end{bmatrix} \end{bmatrix}_{Time} \end{vmatrix} \langle x_U | x_T \rangle$$

Eq 2.8.2: Space-Time Emergence

An implication of this equation, brought out more explicitly through the elaboration of the 'Time' component, is that the layers U, M_1, M_2, and M_3 exist simultaneously. Adding the Light-Matrix derived in Chapter 2.1 enhances Equation 2.8.2 to the Light-Space-Time Emergence form as represented by Equation 2.8.3:

$$Emergence_{light-space-time}$$

$$= \begin{bmatrix} C_\infty: [Pr, Po, K, H] \\ (\downarrow R_{C_K} = R_{C_\infty}) \\ C_K: [S_{Pr}, S_{Po}, S_K, S_H] \\ (\downarrow R_{C_N} = R_{C_K}) \\ C_N: f(S_{Pr} \times S_{Po} \times S_K \times S_H) \\ (\downarrow R_{C_U} = R_{C_N}) \\ C_U: [\leftarrow, |, \rightarrow, \leftrightarrow] \end{bmatrix}_{Light} \begin{bmatrix} M_3 \rightarrow System_X \\ (\uparrow F \rightarrow I) \\ M_2 \rightarrow S_{System_X} \\ (\uparrow Sig \rightarrow F) \\ M_1 \rightarrow Sig_x \\ (\uparrow > P_x) \\ U \rightarrow x_U \end{bmatrix}_{Space} \begin{bmatrix} M_3: All\ t \\ \downarrow \\ M_2: 0 \geq t > \infty \\ \downarrow \\ M_1: 0 > t > \infty \\ \downarrow \\ U \rightarrow \begin{matrix} t \sim E_{Cell}; TC: M_3 \rightarrow U \\ t \sim E_{Human}; TC: U \rightarrow M_3 \end{matrix} \end{bmatrix}_{Time} \begin{matrix} TC \rightarrow x_T \end{matrix} \Big|_{\langle x_U | x_T \rangle}$$

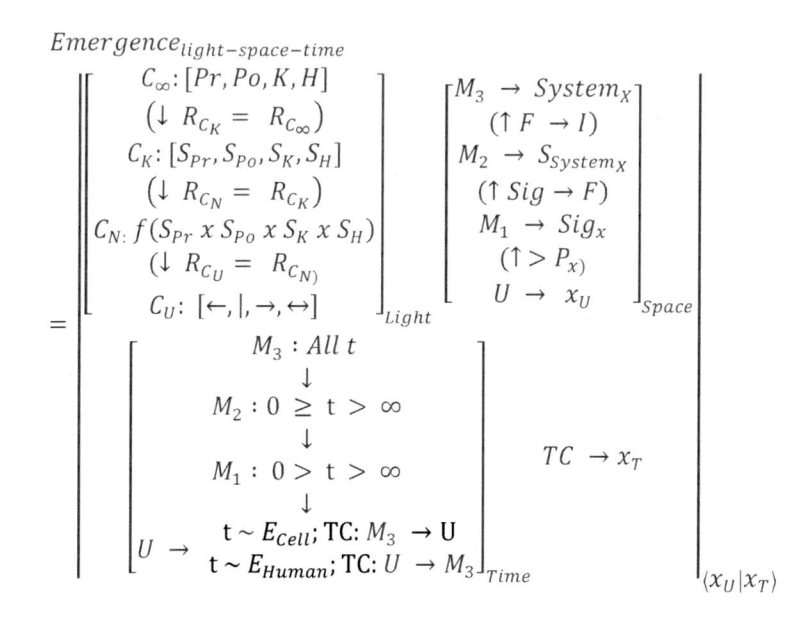

Eq 2.8.3: Light-Space-Time Emergence '

(2.8.3) will be further explored in Sections 3, 4, and 5.

SECTION 3: INTERPRETATION OF QUANTUM PHENOMENA

In light of Sections 1 and 2, this section will offer an interpretation of quantum phenomena. Specifically the following will be reviewed:

- The core hypothesis in interpreting quantum phenomena
- Speed of light and quanta
- Interpreting Schrodinger's Equation
- Interpreting Heisenberg's Uncertainty Principle
- Interpreting other quantum phenomena
- A deeper look at quantization of space, time, matter, and gravity

Chapter 3.1: Interpreting Quantum Phenomenon – A Core Hypothesis

In his book QED: The Strange Theory of Light and Matter (Feynman, 1985), Feynman states that the theory he presents will not explain why or how Nature acts the way it does but will explain with very high accuracy the probability that a photon emitted from a monochromatic light source is detected by a photon detector. Brian Clegg, in his book, The Quantum Age (Clegg, 2014) relates how Feynman in a public lecture about quantum particles says: "You think I am going to explain it to you so you can understand it? No, you are not going to be able to understand it. Why, then, am I going to bother with all this? Why are you going to sit here all this time, when you are not going to be able to understand what I am going to say? It is my task to persuade you not to turn away because you do not understand it. You see, my physics students don't understand it either. This is because I don't understand it. Nobody does."

Feynman jokingly goes on to suggest that the particles route could be absurd – going around Jupiter, to the local hot-dog stand, before reaching the detector. But this absurdity has been generalized and has become an edifice by which quantum nature is now framed and understood.

Physics has, as has all of science, tended to take a solely physical approach to explaining phenomena. Hence it is believed that everything is emergent and therefore in an almost randomized way, as it would likely have to be if the source of causes is conceived as arising from a single and sole layer of reality, as opposed to a possible set of distinct layers. Absurdity has to arise when there is the artificial compression of meta-level action into one level without even recognizing the impetuses that exist from possible meta-levels. Probability functions and uncertainty principles and equations have to be erected when distinctions of action caused by principles of organization that are non-physical in nature are suppressed or ignored. Ignorance can be easily swept under a probability or uncertainty function that is then erected as new knowledge. It is then easy to come to believe that the nature of reality is the nature of the last model that has been proposed to deal with this lack of knowledge, and therefore questionable edifice upon questionable edifice continues to be built often without going back to first principles. No matter how sophisticated or accurate any model may prove to be it will be easier to advance knowledge if it is remembered that any knowledge is itself a model. As Nassim Taleb in the Black Swan (Taleb, 2010) suggests: "You view the world from within a model."

This notion of building edifice upon edifice is somewhat reminiscent of observations made by Joseph Weizenbaum, the MIT computer scientist, in his book Computer Power and Human Reason in writing about his experiments with ELIZA, a natural language processor he had developed (Weizenbaum, 1967). He states: "This reaction to ELIZA showed me more vividly than anything I had seen hitherto the enormously exaggerated attributions an even well-educated audience is capable of making, even strive to make, to a technology it does not understand."

But as is being proposed in the generalized equation for innovation in Chapter 2.6, there is a multi-layered action and meta-levels that have a profound impact on physical phenomena. If once the existence of these meta-layers is admitted, and of the dynamics of organization that prevail for example at M_1, M_2, and M_3, as proposed here, then there are a range of alternative models and theories that can be created.

By way of summary, the following is the core-matrix at the heart of the generalized equation for innovation already derived in Chapter 2.6, which will be referred to in proposing alternative explanations:

$$\begin{bmatrix} M_3 \to System_X \\ (\uparrow F \to I) \\ M_2 \to S_{System_X} \\ (\uparrow Sig \to F) \\ M_1 \to Sig_x \\ (\uparrow > P_{x)} \\ U \to x_U \end{bmatrix}$$

Hence, it is being suggested that just behind the untransformed or physical layer, U, there is a meta-level, M_1, in which the uniqueness of any organization is specified. The specification itself is derived from further meta-levels, M_2 and M_3, which precipitate in infinite ways yet assuring overall system coherence.

This notion of the uniqueness of organization as determined by a meta-level, even in the case of the quantum world, can perhaps be brought home by a thought experiment.

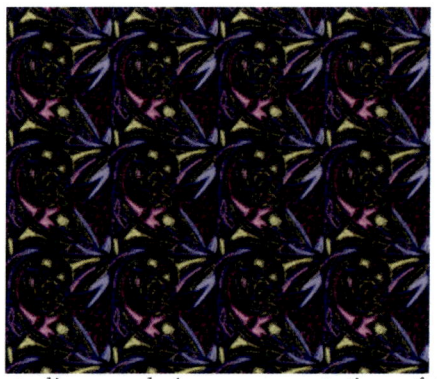

Consider the simplest of atoms – the hydrogen atom. The nucleus comprises one proton around which encircles one electron. Envisioning this atom one can think of a baseball stadium as being representative of the atom. But the central nucleus, in this case the proton, is the size of a baseball. The electron is flying around this baseball at high speed to create an image of the stadium. But one can see in this visualization that the atom mainly comprises empty space. If it is being proposed that the diversity of individual elemental characteristics is determined by the simple possible combinations of protons and electrons and other possible fundamental particles that mainly occupy empty space, then can't it also be proposed that similar structures such as baseball stadiums with a single baseball at the center, or rather all the different structures from the micro to the macro with similar spatial characteristics would also then express different functionality that would then create our reality?

But that then would create complete chaos as one diverse configuration and therefore functionality countered another. This is clearly not the case and therefore form by itself should not determine function. In fact in several design disciplines such as architecture (DeZurko, 1952) and software engineering (Martin, 2002), form is intended to follow function.

Rather keeping the core-matrix in mind perhaps it may be the case that the unique signature of an element, as determined by Sig_x at M_1, precipitates in a self-similar configuration fixed for it by the unique atomic configuration for an element as specified by its atomic weight (number of protons in the nucleus). The unique configuration of an atom as determined by its atomic weight then becomes a "switch" by which the associated function from any possible number of functions as determined by Sig_x precipitates and determines the character of the element.

As things get very small so that limits are approached it stands to reason that if there are indeed multiple layers of organization, something of the influence of the meta-levels should be more directly observable even under conventional techniques. If system models are still strapped to a sole single physical level, then any impetus or influence from a meta-level is therefore likely going to more easily be misinterpreted or misunderstood.

Compensation mathematics and equations will then have to be erected to explain the misunderstandings away.

Chapter 3.2: Speed of Light and Quanta

As discussed conceptually in Section 1 and also mathematically in Chapter 2.1, since c is finite and therefore there is past, present, and future implied by it, this implies that at U a point has to become quanta. This is implicit in the notion of finiteness. Since light takes a finite amount of time to get from A to B, a "unit" of light will require a finite time to traverse that. Quanta at the subatomic level can be thought of as related to this finite time and distance for a unit of light to be expressed.

Planck's discovery that energy at the subatomic level acts as quanta therefore makes sense. It is to be noted though that Planck's treatment of quanta was more as a mathematical convenience that allowed the derivation of an equation that explained the curve of radiation wavelengths at varying temperatures of a heated black-body (Isaacson, 2008). Einstein though postulated quanta as a fundamental property of light itself, rather than as something that arose in the interaction of light with matter as Planck thought. Einstein's theory produced a law of the photoelectric effect where the energy of emitted electrons would depend on the frequency of light. Einstein received the Nobel Prize for this discovery (Isaacson, 2008).

Summarizing, if c is the upper limit of the layer U, then it makes sense that the lower limit h (Planck's constant) should be inversely proportional to c. Hence:

$$h \propto \frac{1}{c}$$

This relationship is substantiated by combining two well-known equations: the first is the electromagnetic equation connecting speed of light with wavelength and frequency, and the second is Einstein's photoelectric equation connecting energy with frequency of light:

(1) $c = \nu\lambda$
(2) $E = h\nu$

Yields:

$$h = \frac{E\lambda}{c}$$

About h, H.A. Lorentz the Dutch scientist has commented in The Science of Nature (Lorentz, 1925): "We have now advanced so far that this constant not only furnishes the basis for explaining the intensity of radiation and the wavelength for which it represents a maximum, but also for interpreting the quantitative relations existing in several other cases among the many physical quantities it determines. I shall mention a few only, namely the specific heat of solids, the photo-chemical effects of light, the orbits of electrons in the atom, the wavelengths of the lines of the spectrum, the frequency of the Roentgen rays which are produced by the impact of electrons of given velocity, the velocity with which gas molecules can rotate, and also the distances between the particles which make up a crystal. It is no exaggeration to say that in our picture of nature nowadays it is the quantum conditions that hold matter together and prevent it from completely losing its energy by radiation."

So just as c sets up the past-present-future experience and reality of U, h suggests that this experience will take place in shells of matter. In the absence of the limit h, as pointed out by Lorentz, only radiation, and no matter would exist.

This reinforces the notion of the four-foldness implicit in the nature of light as already discussed in Section 1.

The suggested variance in the speed of light by meta-layer may also throw some further light on the quantum realm. First, summarizing:

1. At U the speed of light in a vacuum, c_U, is finite at 186,000 miles/sec. This finiteness creates the reality and experience of past-present-future, and further a sense of fragmentation and separation. Further, assuming that c_U is a fundamental upper-limit at U, the inverse of it, $\frac{1}{c_U}$, must define some fundamental lower limit at U. This is indeed the case as Planck's constant, h, is proportional to this. 'h' allows for matter to be sustained, as it fundamentally limits the dispersion of energy as suggested by Lorentz.

2. At M_3, the speed of light, c_{M_3}, is suggested as being ∞ miles/sec. This allows a reality of 'oneness' and the possibility of a suggested four-fold intelligence existing in every point-instant of space-time as suggested in Chapter 1.2.

3. As suggested in Chapter 3.1 the quantum world, here designated by Q, because it is at boundary of U, accesses and interrelates with the meta-levels. As such, the speed of light, c_Q, will appear as a hybrid as in the following figure. Note though that it is really the speed of light at the native or resident layer that becomes active, and that this is simply being represented as c_Q for convenience:

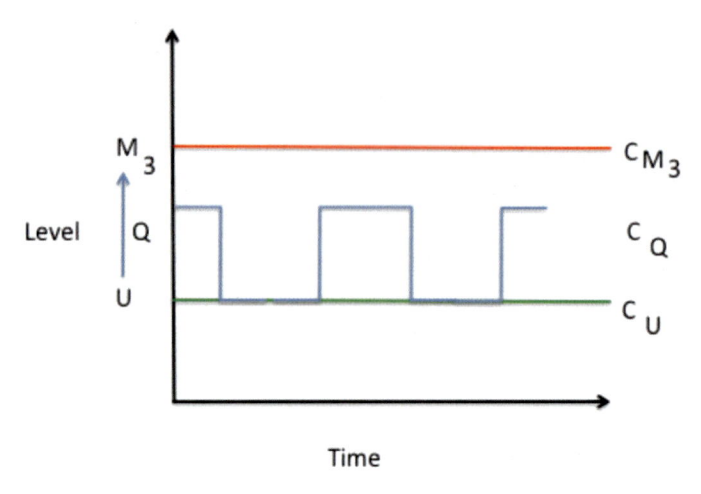

Figure 3.2.1 Speed of Light at Quantum Level

Note that research on the speed of light also indicates that it may go faster than c_U. While the speeds suggested currently through experimental research, and summarized below, may be only incrementally higher than c_U the notion that c_U can be exceeded appears to be put in place:

1. The Heisenberg Uncertainty Principle (to be discussed subsequently in Chapter 3.4) already suggests that photons can travel at any speed, even exceeding c_U, for short periods.
2. Notion of different space-time realities, also known as meta-levels, suggests that light can travel differently in a layer different from the four-dimensional space-time that apparently defines our observable world (Hawking, 1988)
3. In his book QED Feynman (Feynman, 1985) says "...there is also an amplitude for light to go faster (or slower) than the conventional speed of light. You found out in the last lecture that light doesn't go only in straight lines; now, you find out that it doesn't go only at the speed of light! It may

surprise you that there is an amplitude for a photon to go at speeds faster or slower than the conventional speed, c." In research conducted at Humboldt University (Chown, 1990), Scharnhorst has made calculations using the theory of quantum electrodynamics to reveal the possible existence of "faster-than-light" photons. This is known as the Scharnhorst effect.

4. As reviewed in Chapter 2.1, Perkowitz makes the point that the theory of relativity does not disallow particles already moving at c or greater.

The point is that the reality at Q is going to be different than the reality at U. This should be apparent from considering the relation of c_X to the consequent reality as discussed in Section 1. In Q the fundamental lower limit, h, which allows matter to sustain itself, is itself going to fluctuate. Hence, as X tends to M_3, c_X will tend to infinity, and h will become a fraction of itself. As it becomes a fraction of itself the quantization effect will be lowered, and matter will get dispersed more and more easily to in effect take on a wave-like appearance.

99

Chapter 3.3: Interpreting Schrodinger's Equation

Schrodinger's equation, which seeks to model how a quantum state of a quantum system changes with time, or in other words seeks to model matter as a wave rather than as a particle (Stewart, 2012), is depicted in Equation 3.3.1:

$$i\frac{h}{2\pi}\frac{\partial}{\partial x}\psi = \hat{H}\psi$$

Eq 3.3.1: Schrodinger's Equation

ψ depicts a wave form and can be thought of as a probable cloud of possible states. $\hat{H}$ is the Hamiltonian operator which is a focusing function, and in its essence what the equation may be suggesting is that the way a wave form changes over time is equivalent to some expressible state of the possibilities inherent in the cloud of possible states.

But the cloud of possible states is another way of saying that behind the layer U form is represented in another way than at U. If the existence of the meta-levels, and in this case, of Sig_x at M_1, is considered possible, then it is far more reasonable to admit that form is configured by function and the very dynamics of what may appear to have been random, as in many interpretations of quantum phenomena, may now appear to be far more logical. There is now more *context* to interpreting observation at the quantum level.

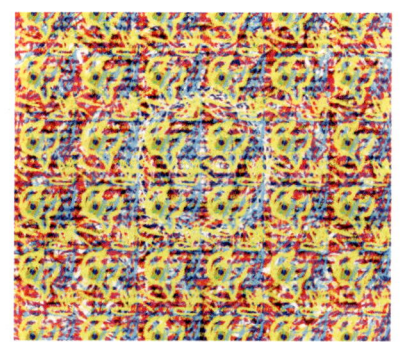

Interestingly Schrodinger himself had misgivings about the applicability of this equation that seemed to apply at the quantum level, to the macro-world (Stewart, 2012). To bring his misgivings to light he invented a thought experiment concerning a cat. This cat would be in a superposed state in a quantum black-box. A radioactive particle, a decaying-particle detector, and a flask of poison, were the other inhabitants of the black-box. At some point the particle will decay, be detected, and as in the thought-experiment at that point, triggered by the decaying particle, the poison in the flask would be released. The cat would then die. But in the meanwhile the cat would be in a superposed states of being both dead and alive. Only when the box was opened would the wave function collapse and a single definite state emerge.

Schrodinger was hoping to highlight the absurdity of the application of having a cat in both a dead and alive state at the macro-level. Instead physicists found this thought experiment to be sensible, even at the macro-level, and began to generalize the findings based on this. Hence, for example the idea of superposition at the physical level began to be thought of as real. It is interesting to note that in his lectures on Schrodinger's equation Feynman (Gottlieb, 2013) has stated "Where did he get that [equation] from? Nowhere. It is not possible to derive it from anything you know. It came out of the mind of Schrödinger".

Considering Schrodinger's equation in the light of the discussion on Q in Chapter 3.2, 'i' is a complex number and suggests the interplay of two dimensions, one being real, and one being 'imaginary'. But the 'imaginary' dimension could be thought of as none other than the meta-levels implicit in this mathematical model, and suggested to be real at Q. Further, $\frac{h}{2\pi}$ is in line with the suggestion just made that h will have to become a fraction of itself as c increases. Hence, the change in the wave function, $\frac{\partial}{\partial x}\psi$, is intimately related to i and $\frac{h}{2\pi}$, and perhaps only makes sense when considered in the context of $i \times \frac{h}{2\pi} \times \frac{\partial}{\partial x}\psi$, which has to be the case when dealing with the integration of dynamics of multiple levels.

Further, the change in the wave function, $\frac{\partial}{\partial x}\psi$, is related to $\hat{H}\psi$, and suggests that there is some system "energy", represented by the Hamiltonian, $\hat{H}$, that when applied to the existing wave, ψ, will indicate how the wave will be expressed going forward.

But as discussed, at Q the dynamics of Sig_X become real, and in fact is a fundamental organizing principle for all organization at U, and starting at the dimension of h.

It may be the case that as the level of complexity of organization at the micro-level increases, as in from quantum, to atomic, to cellular, $\hat{H}$, itself complexifies as it houses further nuances or "superpositions" of Sig_X. Sig_X may be thought of as having more components as the complexity at such micro-levels of organization increases. Further, since it is being proposed here that all complexity or innovation is due to the integration of the dynamics of meta-levels with U, the representation of wave functions or ψ at the boundary world, Q, may be enhanced by consideration of Sig_X. Hence, $\hat{H}$ may be further qualified by notating $\hat{H}_Q$ or $\hat{H}_A$ or $\hat{H}_C$ for quantum, atomic, and cellular respectively, as per the following inequality in Equation 3.3.2:

$\hat{H}_Q < \hat{H}_A < \hat{H}_C$ (where: $\hat{H}_X \propto Energy_{Sig_X}$)

Eq 3.3.2: Hamiltonian Inequality

This inequality is specified by the number of components of Sig_X which it is assumed will have a direct effect on potential and kinetic energy of the wave-system (which are considered to be how $\hat{H}$ is measured) as in Equation 3.3.3:

$$Energy_{Sig_Q} < Energy_{Sig_A} < Energy_{Sig_C}$$

Eq 3.3.3: Quantum, Atomic, Cellular Energy Inequality

Similarly the following wave equations for quantum, atomic, and cellular respectively, as in Equations 3.3.4, 3.3.5, and 3.3.6 may be distinguished:

$$i\frac{h}{2\pi}\frac{\partial}{\partial x}\psi_Q = \hat{H}_Q\psi_Q$$

Eq 3.3.4: Hamiltonian at Quantum Level

$$i\frac{h}{2\pi}\frac{\partial}{\partial x}\psi_A = \hat{H}_A\psi_A$$

Eq 3.3.5: Hamiltonian at Atomic Level

$$i\frac{h}{2\pi}\frac{\partial}{\partial x}\psi_C = \hat{H}_C\psi_C$$

Eq 3.3.6: Hamiltonian at Cellular Level

Chapter 3.4: Interpreting Heisenberg's Uncertainty Principle

In his book, The Little Book of String Theory, Princeton University's Gubser (Gubser, 2010) describes the effect on approaching absolute zero temperature on molecules. He takes the example of water molecules and relates that one cannot make the water molecules colder than absolute zero, -273.15 Celsius, because there is no more thermal energy to suck out at that temperature. However, quantum uncertainty, the phenomenon which relates the momentum and location of electrons in atoms necessitates that the water molecules will still vibrate. Gubser suggests this by considering Heisenberg's uncertainty relation, reproduced in Equation 3.4.1:

$$\Delta p \: X \: \Delta x \geq \frac{h}{4\pi}$$

Eq 3.4.1: Heisenberg's Uncertainty Relation

In Equation 3.4.1, Δp is the uncertainty in a particle's momentum, Δx is the uncertainty in the particle's location, and h is the Planck's constant. In frozen water crystals it is precisely known where the water molecules are, and therefore Δx is fairly small. This means that Δp has to be considerably larger, and therefore that the water molecules are still vibrating even though they are at absolute zero. This innate vibration, known as 'quantum zero-point' energy, expresses the phenomenon of quantum fluctuations.

The Planck's constant order of magnitude (10^{-34}) though, suggests the boundary between U and M_1 and the quantum fluctuations, the uncertainty relation, and the quantum zero-point energy could be an expression of the essential Signature function, Sig_x, that is posited as a key formative force behind organization at U. In this interpretation the thermal energy describes the essential energy at U, while the uncertainty relation may suggest the phenomenon of innovation or function-precipitation "physically" linking M_1 and U. In this case it may be suggested that integration of meta-levels with the surface level, I_U^M, is indicated by the uncertainty relation, as in Equation 3.4.2:

$$I_U^M \rightarrow \Delta p \times \Delta x \geq \frac{h}{4\pi}$$

Eq 3.4.2: Integration of Levels (Leveraging Heisenberg's Uncertainty Relation)

But further, it may also be suggested that the uncertainty principle itself is only valid at U, and that, because of the finiteness of c. This finiteness as already suggested implies h, which implies that if the position of a particle is going to be observed by shining light on it, the light has to have at least a quantum of energy. But to determine the position of a particle accurately, light of a shorter wavelength would have to be used (Hawking, 1988) which would have to have a minimum amount of energy, which in turn would interfere with the velocity and hence momentum of the particle. The uncertainty in measuring the momentum could therefore be thought of as a consequence of the finiteness of the speed of light, c.

If c_U were to approach c_Q though, the quantum would be smaller and the uncertainty in measuring position or momentum would be reduced. At c_{M_3} there would be no uncertainty since light would accurately tell both position and momentum definitively.

Hence, the uncertainty principle may be further qualified, as in Equations 3.4.3, 3.4.4, and 3.4.5:

$$@c_U: \Delta p \times \Delta x \geq \frac{h}{4\pi}$$

Eq 3.4.3: Uncertainty Principle at U

@c_Q: $\Delta p \times \Delta x \rightarrow 0$

Eq 3.4.4: Uncertainty Principle at Q

@c_{M_3}: $\Delta p \times \Delta x = 0$

Eq 3.4.5: Uncertainty Principle at M3

The notion of position and momentum becoming finite at U also may imply that space, time, and quanta are emergent rather than absolute properties, as also suggested in Section 1 and specifically in Chapter 1.6. This is also the conclusion of Arkani-Hamed of the Institute of Advanced Studies in the following thought experiment (Wolchover, 2013):

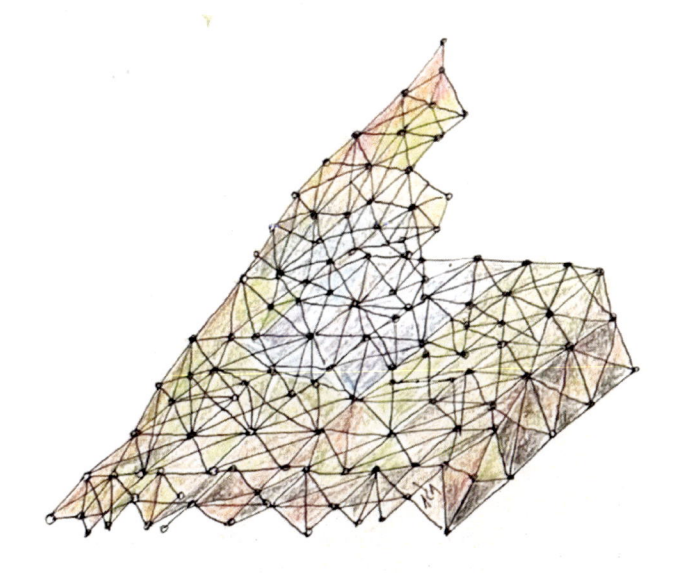

'Locality says that particles interact at points in space-time. But suppose you want to inspect space-time very closely. Probing smaller and smaller distance scales

107

requires ever higher energies, but at a certain scale, called the Planck length, the picture gets blurry: So much energy must be concentrated into such a small region that the energy collapses the region into a black hole, making it impossible to inspect. "There's no way of measuring space and time separations once they are smaller than the Planck length," said Arkani-Hamed. "So we imagine space-time is a continuous thing, but because it's impossible to talk sharply about that thing, then that suggests it must not be fundamental — it must be emergent."

Unitarity says the quantum mechanical probabilities of all possible outcomes of a particle interaction must sum to one. To prove it, one would have to observe the same interaction over and over and count the frequencies of the different outcomes. Doing this to perfect accuracy would require an infinite number of observations using an infinitely large measuring apparatus, but the latter would again cause gravitational collapse into a black hole. In finite regions of the universe, unitarity can therefore only be approximately known.'

Chapter 3.5: Interpreting Other Quantum Phenomena

The following sections suggest alternative explanations to now commonly accepted quantum nature features in the light of the meta-level mathematical model proposed in this book.

Dual wave-particle nature

In Schrodinger's equation the wave aspect, ψ, may actually be an indication that the unique function at the meta-level M_1 as specified by Sig_x, is going to assure itself one way or another, as apparent by the probability distribution of appropriate particles specified by ψ. In other words, whatever particles need to manifest to assure that the meta-level function is fulfilled, will manifest. The wave-particle nature is incomplete without reference to Sig_x at M_1. The wave and the particle are a child of the meta-level function and are in this way of looking at it incidental to what the meta-level function must achieve.

In "The Interpretation of Quantum Mechanics" (Schrodinger, 1995) a series of unpublished papers and talks, published posthumously by Schrodinger's daughter, he suggests that a wave has both a surface and rays, and these move together. The surface suggests the transverse movement of the wave, while the wave-ray the longitudinal movement observed as the particle. They are never separate and exist always together. This suggestion appears to be consistent with the interpretation of wholeness characterized by possible meta-level function that always accompanies the 'observed' or 'incidental' particle(s). In other words, a meta-level math model such as developed in this book suggests that the layers U, M_1, M_2, and M_3 exist simultaneously and the "duality" may be a "quadrality".

Independent states as specified by superposition

Given that emergent phenomena are in reference to a meta-level context, the superposition that Schrodinger's equation suggests does not define and set into motion manifest independent states, but only possibilities that the meta-level function may cause in fulfilling its implicit intent. The notion of multiverses and alternative histories of universe is in this interpretation of Schrodinger's equation, unnecessary. Therefore this may appear to be one of the unnecessary generalizations of quantum behavior at the macro-level that a number of physicists have assumed as true (Saunders et al., 2012).

Quantum tunneling

Quantum tunneling has been proposed as a mechanism by which quantum-sized particles can 'penetrate' boundaries that classical physics says it should not be able to (Clegg, 2014). This ability of a particle to manifest in a region as defined by the wave-function that defines it, rather than by the apparent physical forces of attraction and repulsion that surround it, whether in a cell or the Sun for example, is what is termed quantum tunneling. It has been likened to a quantum teleportation of sorts.

Enzymes, for example, speed up chemical reactions so that processes can be completed orders of magnitude faster inside living cells. Quantum tunneling has been proposed as the way in which this happens so that electrons and protons can vanish from one position in a biomolecule and apparently rematerialize in another without passing through the gap in between (Clegg, 2014).

But suppose there exists a meta-level function, Sig_x, which is the organizational signature of a necessary cellular level energy creation and monitoring function, $Sig_{cellular-energy-type1}$ in this case. In order to fulfill itself it may be suggested that this function which exists just behind any surface visible range, oversees the movement or manifestation of electrons and protons and monitors cellular energy 'type1'. If protons and electrons are the visible sign or precipitation of this fundamental energy required for cell function, they can then be thought of as 'mapping' the path of this meta-level principle of organization. The wave hence depicts the very meta-level principle of energy-organization manifest as the probability that electrons and protons that serve that principle will show up in the locations suggested by the wave.

Al-Khalili expresses in his book on Quantum Biology (Al-Khalili, 2014), that it is difficult to figure out how a packet of energy captured by a cell in the process of photosynthesis actually makes its way so unerringly through chlorophyll molecules to a structure called the reaction centre where its energy is stored. Khalili reports that in an experiment conducted in Berkeley in 2007 laser light was fired at photosynthetic complexes. The research proposed that the energy packets do not hop about randomly, but through quantum effects behave like a spread-out wave, sampling all possible paths, to thereby find the quickest one. But this precisely suggests wholeness as in Schrodinger's own view in his quantum mechanical interpretations (Schrodinger, 1995) and as suggested by the mathematical model for innovation in this dissertation.

Here hence a fine-tuned hypothesis is offered based on the generalized core-matrix. Basically when dealing with limits in nature, and here of the microscopic limits approaching Planck length, l_P, where the structure of space-time is dominated by quantum effects, this is where M_1, precipitates to the visible layer U. Hence many of the phenomena attributed to the physical layer U, are suggested to be the result of an infinite variety of organizational-functions as specified by Sig_x and

113

occurring at M_1. Hence, the integrity of the physical layer, U, is intact, and there is nothing 'weird' or 'absurd' about it. What is being seen or observed though, are the effects of the meta-layers, where 'physics' operates differently, and is modulated by another set of laws. It is perhaps fair to say that Science still has a way to go to completely uncover all those laws.

In his book "What is Life?" (Schrodinger, 1944) Schrödinger suggests that life is based on a principle whereby its macroscopic order is a reflection of quantum-level order, rather than the molecular disorder that characterizes the inanimate world. He called this principle "order from order". He suggested that, unlike inanimate matter, living organisms reach down to the quantum domain and utilize its strange properties in order to operate the extraordinary machinery within living cells. This notion of reaching into or mobilizing a deeper-level order is consistent with the meta-level organizational principles presented here, and summarized in the generalized core-matrix and will be explored in more detail in Sections 4 and 5 of this book.

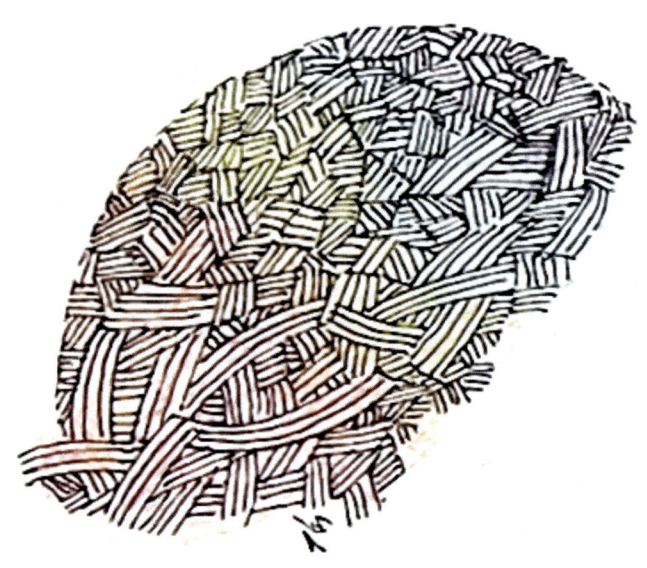

Canceling out of quantum dynamics

In his model on Quantum Electro Dynamics (QED) Feynman suggests that photons do not have to travel in straight lines. In fact, when emitted from a monochromatic light source, they will travel in every direction possible, while arriving at the photon detector. There is a reality of superposition in which all possible paths are traversed by photons. His model allows for combining all the paths together through vector addition to arrive at the path of the straight line recognized by classical physics. In other words, the quantum dynamics cancel themselves out so that one path emerges.

So whether at the micro or macro level vector addition results in a single path to which molecular, atomic, or photonic movement is subject. It is also interesting to note that at the macro-level there is a similar canceling out effect of random molecular motion that yet leaves a containing entity subject to some observed law. For example, if a gas is heated up, inspite of all movement of molecules that cancel one another out, yet the gas will expand in proportion to the applied heat and not in proportion to the apparent molecular motion. Such connection from microscopic behavior to macroscopic properties is the subject of statistical thermodynamics that deals with average properties of the molecules, atoms, or elementary particles in random motion in a system of many such particles (Ebeling & Sokolov, 2005).

These observations are consistent with the idea of a possible existence of an 'organizational function', say 'movement from A to B in an apparent straight line' or the 'equivalence of applied energy', belonging to M_1 as the realm of signatures or organizational functions, to which organizations whether at the photonic, atomic molecular levels at U are subject.

Traveling faster than the speed of light

It has been proposed that information at the quantum level is shared faster than the speed of light (Brumfiel, 2008). But the speed of light is a limit at the physical level, U. At M_1 though, it may be suggested that there exists a general organizational-function, Sig_x, not limited by c, which may be a prime organizing factor in quantum-particle dynamics.

Philosophically, c sets a limit on the ability to transcend U-based space-time. This is suggested by the equations of the alteration of time and space as a body approaches the speed of light. In his book on the special and general theory of relativity (Einstein, 1995), Einstein describes the effect, captured by these equations. Hence, as a body approaches the speed of light, it is perceived by an observer in another frame of reference to be contracting. This contraction – Einstein's Length Contraction - is specified by the factor:

$$Length_{contraction} = Length \left(\sqrt{1 - \frac{v^2}{c^2}} \right)$$

Hence, as the speed of an object, v, increases, the perceived body contracts in dimension. At v = c, the body basically disappears. In other words it can be thought of as having transcended the space continuum, or broken into another space-reality specific to the meta-levels. The length contraction, suggested therefore as being specific to U, is suggested by Equation 3.5.1. Recall that U, in the mathematical model expressed in this book, refers to the untransformed or visible layer which is subject to dynamics not only at the level of U, but also from each of the meta-layers, M_1 through M_3:

$$Length_{contraction-U} = Length_U \left(\sqrt{1 - \frac{v_U^2}{c_U^2}} \right)$$

116

Eq 3.5.1: Length Contraction at Layer U (Leveraging Einstein Theory of Relativity)

Similarly, as a body approaches the speed of light, as perceived by an observer in a 'stationary' frame of reference, there is a time elongation, so that time moves much slower in the moving frame. This is specified at U by the following equation, Equation 3.5.2:

$$Time_{elongation-U} = \frac{Time_U}{\sqrt{1 - \frac{v_U^2}{c_U^2}}}$$

Eq 3.5.2: Time Elongation at U (Leveraging Einstein Theory of Relativity)

Here too, as v approaches c, time elongates representing a slow down in time at U, also similarly indicating the urge to transcend time as it is experienced at U.

From these equations it is also clear that space and time vary together. c is constant while time and space can vary. This also reinforces the notion as suggested in Section 1 about Light creating the 'context' for operation at U. Hence it is that c sets a limit on the ability to transcend space-time.

In considering operations at U, it can be summarized that while time and space can vary based on velocity of a body, or in other words space-time experience can change relative to each body or organization at U, c sets the limit to how much the experience of space-time can vary.

Entanglement

This quantum property has been invoked in describing how birds navigate across the earth (Al-Khalili, 2014). Assume though that there is an organizational-function Sig_x where x = 'flock migration in winter', for example.

This is an M_1 dynamic and particular organizations at the physical layer are subject to it. Quantum particles may or may not have to be involved in this. As Khalili relates though, 'studies of the European robin suggest that it has an internal chemical compass that utilizes entanglement. This phenomenon describes how two separated particles can remain instantaneously connected via a quantum link. The current best guess is that this takes place inside a protein in the bird's eye, where quantum entanglement makes a pair of electrons highly sensitive to the angle of orientation of the Earth's magnetic field, allowing the bird to "see" which way it needs to fly.'

But also there have been experiments where a photon of light is used to create two entangled photons (Vivoli, 2016). These then share properties when separated. Once entangled, by whatever mechanism, it could be suggested that it is an organizational-function such as Sig_x that takes over. Space-time constraints are therefore changed, and a 'quantum' link is now in effect.

Going backward in time

In his book QED: The Strange Theory of Light and Matter, Feynman (Feynman, 1985) suggests that there are particles that can move backward in time as suggested in Figure 3.5.1.

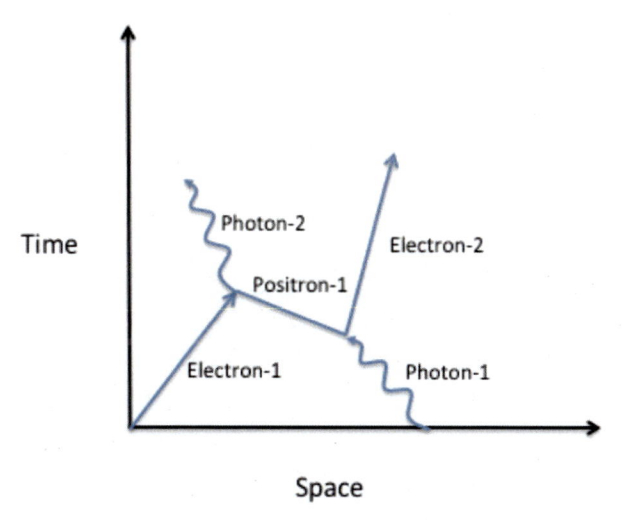

Figure 3.5.1 Particle Moving Backward in Time (Feynman, 1985)

Illustrated is an electron, Electron-1, and a photon, Photon-1, moving towards one another. At some time Electron-1 decays into Photon-2, and Positron-1 (an electron with a positive charge) which then moves backward in time. It then appears to interact with Photon-1, and a new Electron-2 is created.

If however, the existence of a meta-level is assumed, as per the mathematical model developed in this book then spontaneous particle generation from a meta-level organizational-function (in this case organizational-function = "light-energy conversion") could remove the notion of particles moving backward in time. As Stephen Hawking has said in A Brief History of Time (Hawking, 1988), if backward time travel were possible, why aren't there any visitors from the future?

Alternatively, as suggested by the discussion in Chapter 1.2 and 3.2, if the speed of light were to approach infinite, then the whole notion of time changes so that past, present, and future, are blurred.

Summary

So it appears to be possible, at least from the qualitative discussions and inferred inductions from literature, to interpret these afore-mentioned quantum effects – superposition and existence of multiverses, dual-wave particle nature, quantum tunneling, traveling faster than the speed of light, entanglement, going backward in time, using an alternative model as presented here.

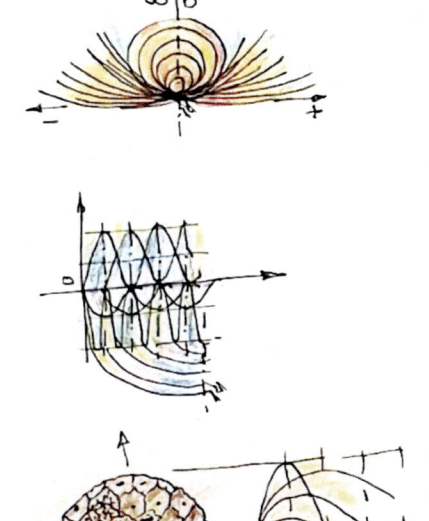

The use of the alternative model allows physical reality to maintain its integrity regardless of scale, since the "weirdness" – superposition, dual-wave particle nature, etc. - experienced at the quantum level is possible through admitting the existence of meta-layers with a different "physics" existing at each meta-layer.

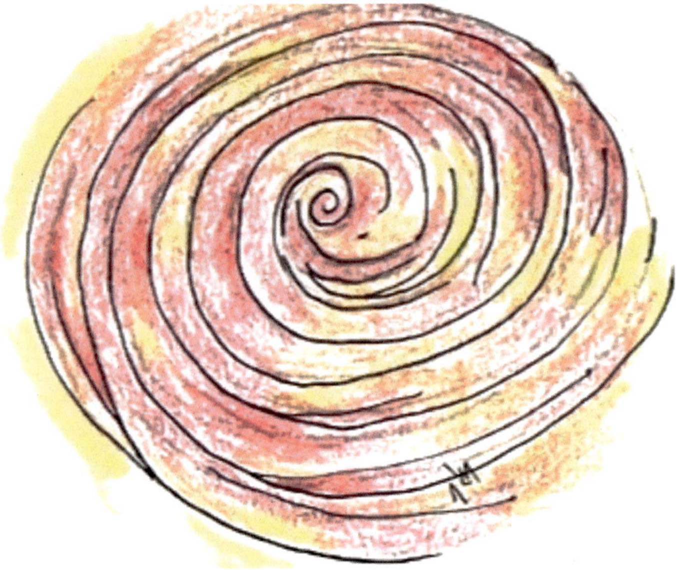

Chapter 3.6: A Deeper Look at Quantization of Space, Time, Matter, and Gravity

The Light-Space-Time Emergence equation (2.8.3) derived in Chapter 2.8, and reproduced here for convenience, offers some insight into the process of quantization. Reproducing the equation:

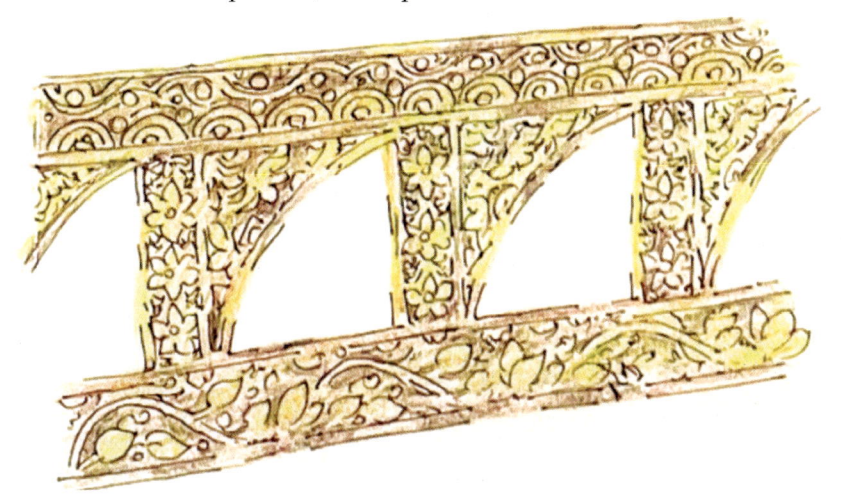

123

$$Emergence_{light-space-time}$$

$$= \begin{bmatrix} \begin{bmatrix} C_\infty: [Pr, Po, K, H] \\ (\downarrow R_{C_K} = R_{C_\infty}) \\ C_K: [S_{Pr}, S_{Po}, S_K, S_H] \\ (\downarrow R_{C_N} = R_{C_K}) \\ C_{N:} f(S_{Pr} \times S_{Po} \times S_K \times S_H) \\ (\downarrow R_{C_U} = R_{C_N}) \\ C_U: [\leftarrow, |, \rightarrow, \leftrightarrow] \end{bmatrix}_{Light} \begin{bmatrix} M_3 \rightarrow System_X \\ (\uparrow F \rightarrow I) \\ M_2 \rightarrow S_{System_X} \\ (\uparrow Sig \rightarrow F) \\ M_1 \rightarrow Sig_x \\ (\uparrow > P_{x)} \\ U \rightarrow x_U \end{bmatrix}_{Space} \\ \begin{bmatrix} M_3 : All\ t \\ \downarrow \\ M_2 : 0 \geq t > \infty \\ \downarrow \\ M_1 : 0 > t > \infty \\ \downarrow \\ U \rightarrow \begin{matrix} t \sim E_{Cell}; TC: M_3 \rightarrow U \\ t \sim E_{Human}; TC: U \rightarrow M_3 \end{matrix} \end{bmatrix}_{Time} \quad TC \rightarrow x_T \end{bmatrix} \langle x_U | x_T \rangle$$

The Light-Matrix suggests that there are particular kinds of quantization that could occur. Along the vertical realm these can be thought of as inter-relating one layer of the matrix with the previous layers. Hence the kind of quantization that is relevant to the physical layer, U, would relate c_U with c_x. This can be represented by h_U, where h stands for Planck's constant. Note that in this model there are other fundamental quantization possible that may inter-relate a specific light-layer with others above it.

In looking at the Light-Matrix it is also clear that each subsequent layer below the top layer, describing the reality set up by an infinite speed, is emergent. Hence space, time, matter, and gravity, can be thought of as emergent phenomena. The emergence itself can be thought of as a function of Light-Matrix. At layer U, and as discussed in Chapter 1.6, Space is related to Light's property of Knowledge, Time is related to Light's property of Power, Matter or Energy is related to Light's property of Presence, and Gravity is related to Light's Property of Harmony or Nurturing. Each of these will emerge in a particular way and it is possible that there are multiple h's.

Hence, the Planck's constant for matter or energy, which we are already familiar with from past discoveries in

science, can be depicted as h_{UPr}, since it is related to Presence (Pr). But similarly, quantization for space, time, and gravity could potentially be governed by other similar constants, referred to as h_{UK}, h_{UP}, and h_{UH}, and related to Knowledge (K), Power (P), and Harmony (H), respectively. Since the relationship between these constants, in absolute terms, is uncertain, this can be represented by the equality-inequality as depicted by Equation 3.6.1:

$$h_{UPr} \leq\geq h_{UK} \leq\geq h_{UP} \leq\geq h_{UH}$$

Eq 3.6.1: Equality-Inequality Relationship Between Different 'Planck' Constants

Further, the quantization-window, as it were, potentially

allows the precipitation of, or inter-relation with, or creation of a cohesive and compelling meta-function or signature as modeled by the c_N/M_1 layer. This quantization-window is positioned as being key in allowing a phenomenon of quantum-certainty to occur and will be explored in greater detail in the subsequent chapters on the making of history. The 'Time' matrix, in (2.8.3) reproduced above, indicates the general default direction under which such quantum-certainty can occur at Layer U. Generally at the pre-human level it may proceed more automatically by the dynamics of the system itself. Beyond this level it will generally occur through the action of a cohesive will. The opening of this quantization-window is modeled by the 'Space' matrix and will happen when habitual patterns are overcome so that 'will' becomes cohesive.

The specific quantization that are occurring along the space, time, energy/matter, and gravity dimensions are modeled by the following equations, that are all based on the Sig_x equations derived in Chapter 2.4. As a reminder (2.4.5) is reproduced here for convenience:

$$Sig = Xa + \overline{Yb_{0-n}}$$

$$where: \begin{bmatrix} X \in [S_{System_{Pr}}, S_{System_P}, S_{System_K}, S_{System_N}] \\ Y \in [S_{System_{Pr}}, S_{System_P}, S_{System_K}, S_{System_N}] \\ a, b \text{ are integers}; a > b \end{bmatrix}$$

The Structure of Space

The structure of Space, which holds the seeds of knowledge of all that will emerge, hence, is modeled in the following way as specified by Equation 3.6.2:

$$Space_{quantization} = h_{UK}(Xa + \overline{Yb_{0-n}})$$

$$where: \begin{bmatrix} X \in [S_{System_K}] \\ Y \in [S_{System_{Pr}}, S_{System_P}, S_{System_K}, S_{System_N}] \\ a, b \text{ are integers}; a > b \end{bmatrix}$$

Eq 3.6.2: Space Quantization

In this model 'space' as an emergent property of Light, is structured by infinite seeds of knowledge. But because the emergence is taking place in a layer of reality generally itself structured by a finite speed of light, c, it has to be quantized. The quantization assures that the knowledge is not dissipated, but can accumulate to create seeds, and therefore, the structure of space itself. It is assumed that there is some kind of 'Planck's constant' in effect, that is modeled as being specific to the way knowledge or space may be quantized – hence, h_{UK}.

The Structure of Time

The structure of Time, which holds the inevitability of the seeds or knowledge emerging in a phased maturity, hence, is modeled in the following way as specified by Equation 3.6.3:

$$Time_{quantization} = h_{UP}(Xa + \overline{Yb_{0-n}})$$

$$\text{where:} \begin{bmatrix} X \in [S_{System_P}] \\ Y \in [S_{System_{Pr}}, S_{System_P}, S_{System_K}, S_{System_N}] \\ a, b \text{ are integers}; a > b \end{bmatrix}$$

Eq 3.6.3: Time Quantization

In this model 'time' as an emergent property of Light, is structured by an inevitable process of maturity, which due to its inevitability, is related to power. But because the emergence is taking place in a layer of reality generally itself structured by a finite speed of light, c, it has to be quantized. The quantization assures that the power is not dissipated, but can accumulate to express phased maturity, and therefore, the structure of time itself. It is assumed that there is some kind of 'Planck's constant' in effect, that is modeled as being specific to the way power or time may be quantized – hence, h_{UP}.

Energy & Matter

Energy, which through a process of containment can result in Matter, hence, is modeled in the following way as specified by Equation 3.6.4:

$$Energy_{quantization} = h_{UPr}(Xa + \overline{Yb_{0-n}})$$

$$where: \begin{bmatrix} X \in [S_{System_{Pr}}] \\ Y \in [S_{System_{Pr}}, S_{System_P}, S_{System_K}, S_{System_N}] \\ a, b \text{ are integers}; a > b \end{bmatrix}$$

Eq 3.6.4: Energy Quantization

In this model 'energy' as an emergent property of Light, results in the reality of matter. But because the emergence is taking place in a layer of reality generally itself structured by a finite speed of light, c, it has to be quantized. The quantization assures that the energy is not dissipated, but can accumulate to create matter. Planck's constant is referred to as - h_{UPr}.

Gravity

Gravity, which holds seemingly distinct objects in the layer of reality created by c together in a harmony, hence, is modeled in the following way as specified by Equation 3.6.5:

$$Gravity_{quantization} = h_{UH}(Xa + \overline{Yb_{0-n}})$$

$$where: \begin{bmatrix} X \in [S_{System_H}] \\ Y \in [S_{System_{Pr}}, S_{System_P}, S_{System_K}, S_{System_N}] \\ a, b \text{ are integers}; a > b \end{bmatrix}$$

Eq 3.6.5: Gravity Quantization

In this model 'gravity' as an emergent property of Light, results in a harmonious collectivity of seemingly independent objects. But because the emergence is taking place in a layer of reality generally itself structured by a finite speed of light, c, it has to be quantized. The quantization assures that the harmony is not dissipated, but can accumulate to express more and more complex collectivities on large-scale. It is assumed that there is some kind of 'Planck's constant' in effect, that is modeled as being specific to the way harmony or time may be quantized – hence, h_{UH}.

SECTION 4: QUANTUM CERTAINTY & NATURAL HISTORY

This section explores the process of quantum certainty in determining natural history.

First the structure of the cell will be explored. This will be followed by an example to illustrate the process of quantum certainty in determining how natural history may play out.

132

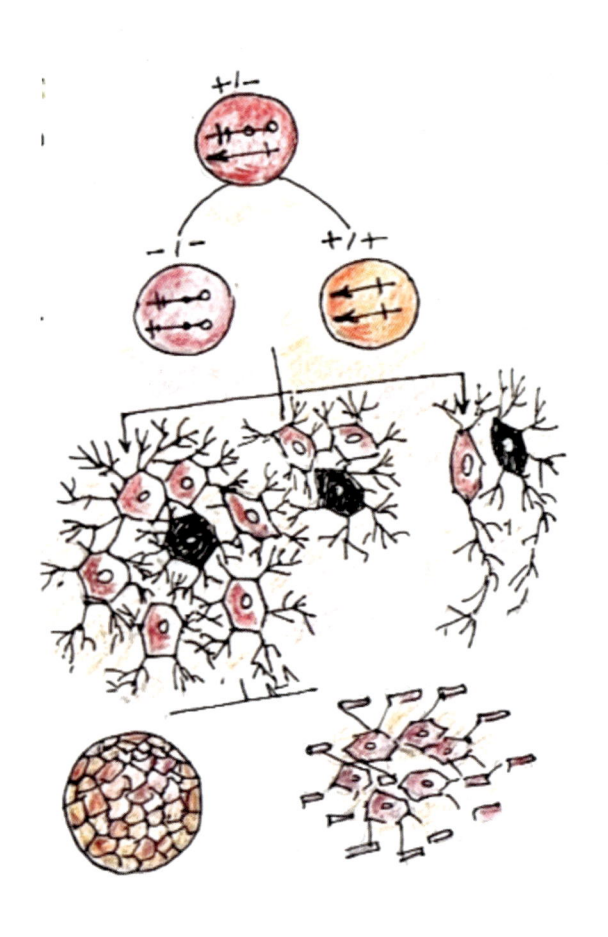

Chapter 4.1: Living Cells as the Fulcrum of Natural History

The living cell has a universe of adaptability embedded in it. In 'The Machinery of Life', Goodsell, an Associate Professor of Molecular Biology at the Scripps Research Institute (Goodsell, 2010) suggests that every living thing on Earth uses a similar set of molecules to eat, to breathe, to move, and to reproduce. There are molecular machines that do the myriad things that distinguish living organisms that are identical in all living cells. This nanoscale machinery of cells uses four basic molecular plans with unique chemical personalities: nucleic acids, proteins, lipids, and polysaccharides.

But further, it is a combination of these molecular plans that can cause specialization of a cell to create any feature of a plant or animal. At the level of plants for example, striking colors and scents can become the natural features of a plant. It can adapt its shape to allow various insects or bees to enter into it in different ways. It can store a myriad of nutrients in a myriad different ways. It can develop poisons to keep predatory animals away. At the level of animals for example, skins can toughen into hard shells. Feet can develop suction caps to allow mobility in a vertical direction. Eyes can develop broad vision, telescopic vision, or microscopic vision. The mechanics of all adaptability can be traced to the ability of the four molecular plans to coordinate and generate more granular cell-level features that give it some evolutionary advantage in the scheme of things. All Natural History, hence, can be traced to this ability of cells to specialize along a diverse array of specialized lines.

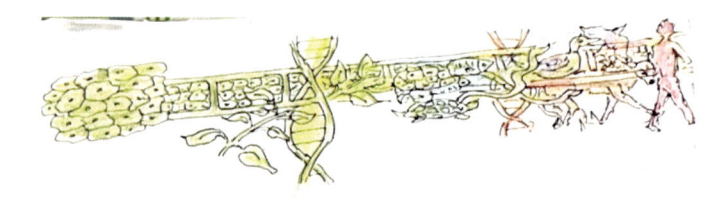

In this chapter we will examine the mathematics of the molecular plans that form the bases for such specialization to occur.

134

Nucleic Acids

Nucleic acids basically encode information. They store and transmit the genome, the hereditary information needed to keep the cell alive. They function as the cell's librarians and contain information on how to make proteins and when to make them.

They are hence, the keepers of a cell's knowledge, its wisdom, its ability to make laws, the vehicle to spread knowledge within cells and to the next generation of cells. Being so, one can see that there is similarity with the set for system-knowledge highlighted earlier in Chapter 2.3.

Reproducing Equation 2.3.3:

S_{System_K}
∋ $[Wisdom, Law\ Making, Spread\ of\ Knowledge\ ...]$

Nucleic acids can therefore be thought of as a precipitation of system-knowledge at the cellular level.

Hence, a nucleic acid will have a generalized signature, as in Equation 4.1.1, derived from the system-knowledge family:

$$Sig_{nucleic\ acid} = Xa + \overline{Yb_{0-n}}\ where \begin{bmatrix} X \in [S_{System_K}] \\ Y \in [S_{System_{Pr}}, S_{System_P}, S_{System_K}, S_{System_N}] \\ a, b\ are\ integers; a > b \end{bmatrix}$$

Eq 4.1.1: Nucleic Acid

The primary element X could be an attribute or function such as 'keeper of knowledge'. Secondary elements Y could be 'protein laws', 'generational knowledge', amongst others. The collectivity of elements as per the equation would specify the character of a nucleic acid.

DNA and RNA, two types of nucleic acids, would hence have the equations as specified by Equation 4.1.2 and 4.1.3 respectively.

$$Sig_{DNA} = \frac{Xa +}{Yb_{0-n}} \text{ where } \begin{bmatrix} X \in [S_{System_K}] \\ Y \in [S_{System_{Pr}}, S_{System_P}, S_{System_K}, S_{System_N}] \\ a, b \text{ are integers}; a > b \end{bmatrix}$$

Eq 4.1.2: DNA

$$Sig_{RNA} = \frac{Xa +}{Yb_{0-n}} \text{ where } \begin{bmatrix} X \in [S_{System_K}] \\ Y \in [S_{System_{Pr}}, S_{System_P}, S_{System_K}, S_{System_N}] \\ a, b \text{ are integers}; a > b \end{bmatrix}$$

Eq 4.1.3: RNA

The primary element X in (4.1.2) and (4.1.3) would be the same as that for nucleic acids (4.1.1). The secondary elements Y however will be a larger and more specific set with many elements in common with (4.1.1).

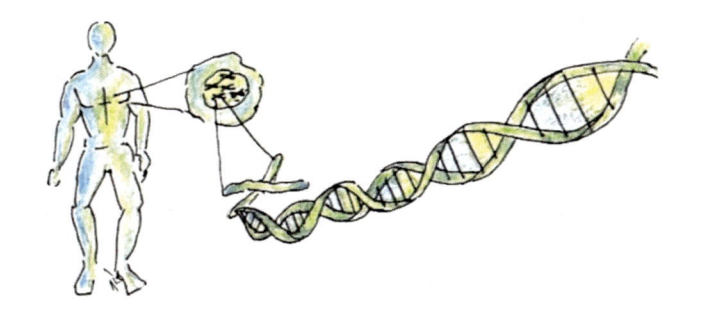

Proteins

Proteins are the cells work-horses. Look anywhere in a cell and one will see proteins at work. Proteins are built in thousands of shapes and sizes, each performing a different function. As Goodsell describes, "some are built simply to adopt a defined shape, assembling into rods, nets, hollow spheres, and tubes. Some are molecular motors, using energy to rotate, or flex, or crawl. Many are chemical catalysts that perform chemical reactions atom-by-atom, transferring and transforming chemical groups exactly as needed." With their wide potential for diversity, proteins are constructed to perform most of the everyday tasks of the cells. In fact human cells build around 30,000 different kinds of proteins to execute on the diverse array of cellular level tasks.

Proteins hence, exist for service, to bring about perfection at the level of the cell, are characterized by extreme diligence and perseverance, and so on. Being so, one can see that there is similarity with the set for system-presence highlighted earlier in Chapter 2.3. Reproducing Equation 2.3.1:

$$S_{System_{Pr}} \ni [Service, Perfection, Diligence, Perseverance, ...]$$

Proteins can therefore be thought of as a precipitation of system-presence at the cellular level.

Hence, a protein could have a generalized signature, as in Equation 4.1.4, derived from the system-presence family:

$$Sig_{protein} = \frac{Xa+}{Yb_{0-n}} \quad where \quad \begin{bmatrix} X \in [S_{System_{Pr}}] \\ Y \in [S_{System_{Pr}}, S_{System_P}, S_{System_K}, S_{System_N}] \\ a, b \text{ are integers}; a > b \end{bmatrix}$$

Eq 4.1.4: Protein

This could yield a vast number of functional proteins. In fact it may be possible that the 30,000 or so known proteins created by the human cell could each be specified by a signature equation of this nature. It may be possible to map existing proteins to functionality as suggested by the four sets of molecular plans.

Consider Insulin, for example. Insulin regulates the metabolism of carbohydrates, fats and protein by promoting the absorption of, especially, glucose from the

138

blood into fat, liver and skeletal muscle cells. Equation 4.1.5 for Insulin would hence be:

$$Sig_{insulin} = Xa + \overline{Yb_{0-n}} \text{ where } \begin{bmatrix} X \in [S_{System_{Pr}}] \\ Y \in [S_{System_{Pr}}, S_{System_P}, S_{System_K}, S_{System_N}] \\ a, b \text{ are integers}; a > b \end{bmatrix}$$

Eq 4.1.5: Insulin

The primary element X could be an attribute or function such as 'workhorse. Secondary elements Y could be 'metabolic regulation, 'glucose absorption', 'blood to fat channel', amongst others. The collectivity of elements as per the equation would specify the character of insulin.

Consider Histones as another example. They are the chief protein components of chromatin, acting as spools around which DNA winds, and playing a role in gene regulation. Without histones, the unwound DNA in chromosomes would be very long (a length to width ratio of more than 10 million to 1 in human DNA). Equation 4.1.6 for Histones would hence be:

$$Sig_{histones} = Xa + \overline{Yb_{0-n}} \text{ where } \begin{bmatrix} X \in [S_{System_{Pr}}] \\ Y \in [S_{System_{Pr}}, S_{System_P}, S_{System_K}, S_{System_N}] \\ a, b \text{ are integers}; a > b \end{bmatrix}$$

Eq 4.1.6: Histones

The secondary element Y would have elements such as 'gene regulation', 'unwound DNA management', amongst others.

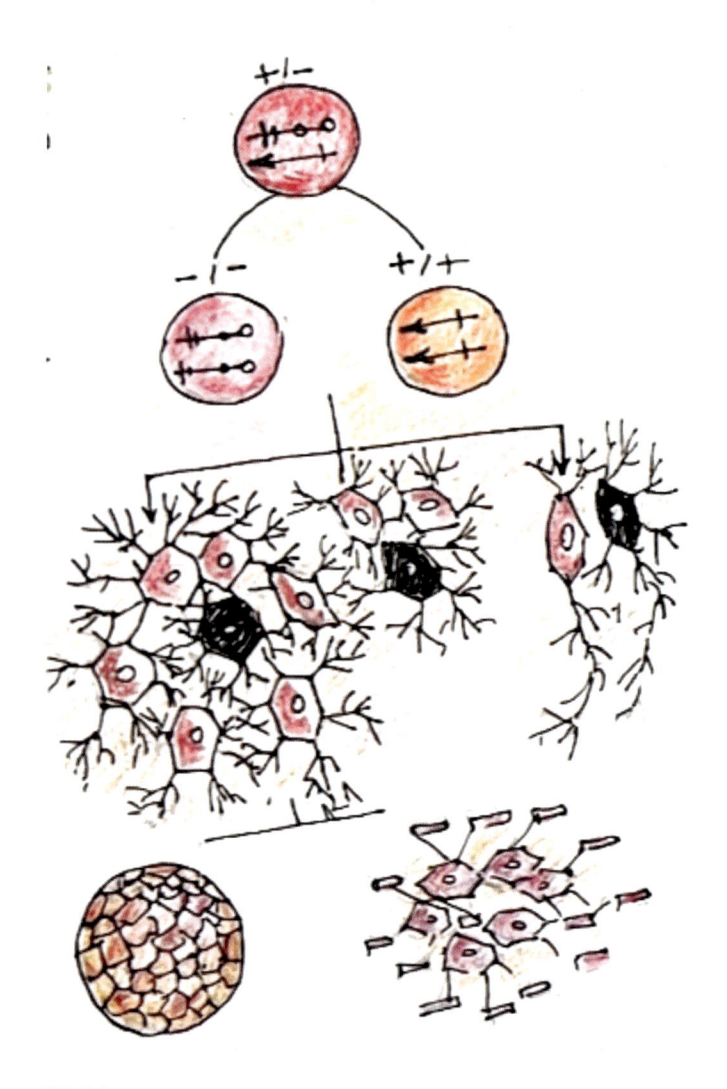

Lipids

Lipids by themselves are tiny molecules, but when grouped together form the largest structures of the cell. When placed in water lipid molecules aggregate to form huge waterproof sheets. These sheets easily form boundaries at multiple levels and allow concentrated

140

interactions and work to be performed within a cell. Hence, the nucleus and the mitochondria are contained within lipid-defined compartments. Similarly, each cell itself is contained within a lipid-defined boundary.

Lipids are therefore promoters of relationship, of harmony in the cell, of nurturing the cell-level division of labor, of allowing specialization and uniqueness to emerge, hence perhaps of earlier forms of compassion and love, and so on. The notion of such early forms of compassion is consistent with the biologist's perspective that at some point a gene for compassion was developed in pre-human species (Wright, 2009). Being so, one can see that there is similarity with the set for system-nurturing highlighted earlier in Chapter 2.3. Reproducing Equation 2.3.4:

S_{System_N}
$\ni [Love, Compassion, Harmony, Relationship \ ...]$

This function of harmonization suggests that lipids can therefore be thought of as a precipitation of system-nurturing at the cellular level.

Lipids could have a generalized signature, as in Equation 4.1.7, derived from the system-nurturing family:

$$Sig_{lipid} = Xa + \overline{Yb_{0-n}} \ where \ \begin{bmatrix} X \in [S_{System_N}] \\ Y \in [S_{System_{Pr}}, S_{System_P}, S_{System_K}, S_{System_N}] \\ a, b \ are \ integers; a > b \end{bmatrix}$$

Eq 4.1.7: Lipid

Specific lipids such as monoglycerides and phospholipids cold have the following equations:

$$Sig_{monoglyceride} = Xa + \overline{Yb_{0-n}} \; where \begin{bmatrix} X \in [S_{System_N}] \\ Y \in [S_{System_{Pr}}, S_{System_P}, S_{System_K}, S_{System_N}] \\ a, b \; are \; integers; a > b \end{bmatrix}$$

Eq 4.1.8: Monoglyceride

$$Sig_{phospholipids} = Xa + \overline{Yb_{0-n}} \; where \begin{bmatrix} X \in [S_{System_N}] \\ Y \in [S_{System_{Pr}}, S_{System_P}, S_{System_K}, S_{System_N}] \\ a, b \; are \; integers; a > b \end{bmatrix}$$

Eq 4.1.9: Phospholipids

The primary element X shared by each of the lipids could be an attribute or function such as 'compartmentalization'. Secondary elements Y could be of the nature of 'work breakdown', 'intra-cell love', amongst others, and would vary with each different kind of lipid.

Polysaccharides

Polysaccharides are long, often branched chains of sugar molecules. Sugars are covered with hydroxyl groups, which associate to form storage containers. As a result polysaccharides function as the storehouse of cell's energy. In addition polysaccharides are also used to build some of the most durable biological structures. The stiff shell of insects, for example are made of long polysaccharides.

Polysaccharides function to create energy, power, courage, strength thereby readying the cell for adventure, and so on. Being so, one can see that there is similarity with the set for system-power highlighted previously in Chapter 2.3. Reproducing Equation 2.3.2:

$$S_{System_P} \ni [Power, Courage, Adventure, Justice, ...]$$

Providing energy and strength, polysaccharides can be thought of as a precipitation of system-power at the cellular level.

Polysaccharides could have a generalized signature, as in Equation 4.1.10, derived from the system-power family:

$$Sig_{polysaccharide} = Xa + \overline{Yb_{0-n}} \text{ where } \begin{bmatrix} X \in [S_{System_P}] \\ Y \in [S_{System_{Pr}}, S_{System_P}, S_{System_K}, S_{System_N}] \\ a, b \text{ are integers}; a > b \end{bmatrix}$$

Eq 4.1.10: Polysaccharide

Glycogen is an example of a polysaccharide. Glycogen forms an energy reserve that can be quickly mobilized to meet a sudden need for glucose, but one that is less compact and more immediately available as an energy reserve than say triglycerides. Equation 4.1.11 for Glycogen follows:

$$Sig_{glycogen} = Xa + \overline{Yb_{0-n}} \text{ where } \begin{bmatrix} X \in [S_{System_P}] \\ Y \in [S_{System_{Pr}}, S_{System_P}, S_{System_K}, S_{System_N}] \\ a, b \text{ are integers}; a > b \end{bmatrix}$$

Eq 4.1.11: Glycogen

Cellulose is another example of a polysaccharide. Cellulose is a polymer made with repeated glucose units bonded together by beta-linkages. Humans and many animals lack an enzyme to break the beta-linkages, so

144

they do not digest cellulose. Equation 4.1.12 for Cellulose follows:

$$Sig_{cellulose} = Xa + \overline{Yb_{0-n}} \quad where \begin{bmatrix} X \in [S_{System_P}] \\ Y \in [S_{System_{Pr}}, S_{System_P}, S_{System_K}, S_{System_N}] \\ a, b \text{ are integers}; a > b \end{bmatrix}$$

Eq 4.1.12: Cellulose

The primary element for the preceding polysaccharides X would be along the lines of 'energy storage'. The secondary elements may vary, with a Y element for Glycogen being 'rapid energy deployment' for example, and a Y element for Cellulose being 'bonded energy', for example.

Chapter 4.2: Quantum-Certainty and the Adaptabilities of the Cell

The quantum is a window into layers of reality behind the surface layer U. This insight was captured by Schrodinger's Equation and by Heisenberg's Uncertainty Principle, the former suggesting that matter, or what is going to appear materially, is a function of some incredible number of superposed states containing infinite possibility, and the latter suggesting that quantum fluctuation due to an always buzzing pregnant-infinity, lies behind everything even when seemingly still.

The Light-Space-Time Emergence equation (2.8.3) derived in Chapter 2.8, and reproduced here for convenience, may provide some insight into the dynamics of the pregnant-infinity, and of quantum-certainty in the adaptabilities of the cell. Reproducing the equation:

$$Emergence_{light-space-time}$$

$$= \begin{Vmatrix} \begin{bmatrix} C_\infty: [Pr, Po, K, H] \\ (\downarrow R_{C_K} = R_{C_\infty}) \\ C_K: [S_{Pr}, S_{Po}, S_K, S_H] \\ (\downarrow R_{C_N} = R_{C_K}) \\ C_{N:} f(S_{Pr} \times S_{Po} \times S_K \times S_H) \\ (\downarrow R_{C_U} = R_{C_N)} \\ C_U: [\leftarrow, |, \rightarrow, \leftrightarrow] \end{bmatrix}_{Light} \begin{bmatrix} M_3 \rightarrow System_X \\ (\uparrow F \rightarrow I) \\ M_2 \rightarrow S_{System_X} \\ (\uparrow Sig \rightarrow F) \\ M_1 \rightarrow Sig_x \\ (\uparrow > P_{x)} \\ U \rightarrow x_U \end{bmatrix}_{Space} \\ \begin{bmatrix} M_3 : All\ t \\ \downarrow \\ M_2 : 0 \geq t > \infty \\ \downarrow \\ U \rightarrow \begin{matrix} M_1 : 0 > t > \infty \\ \downarrow \\ t \sim E_{Cell}; TC: M_3 \rightarrow U \\ t \sim E_{Human}; TC: U \rightarrow M_3 \end{matrix} \end{bmatrix}_{Time} TC \rightarrow x_T \end{Vmatrix} \langle x_U | x_T \rangle$$

The Light-Matrix, the top-left matrix in (2.8.3), suggests the possibilities in the pregnant-infinity. These possibilities, recall, have been set up by realities so created by light traveling at different speeds. While prevalent dynamics at the surface layer, U, are always influenced by dynamics from the deeper layers, there is a state that can be created that will allow a more focused and intentional activation of the meta-functions resident in the deeper layers. Such activation will potentially trigger a series of processes culminating in a state of quantum-certainty as will be explored through the rest of this chapter.

The Space-Matrix, the top-right matrix in (2.8.3), suggests the dynamics involved in creating such an activation-state. Essentially this involves the overcoming of patterns at U. Consider the case of a hypothetical insect, for example, that is always prey to other predators. At some point a visceral urge to overcome some of these predators may arise in the insect. This visceral urge is a breaking of habitual patterns common to that species of insect. If it is deep enough and pervasive enough, it can be thought of as an activation-state and will allow the creation of a quantum-window so that at every micro level there is now a connection with the quantum worlds, Q, behind. It is proposed that for effective adaptability to come about it is only such a pervasive state that will open a quantum-window to effectively stimulate the interaction between layers that will ultimately allow adaptability to occur. This activation-state can be further specified by leveraging (2.7.4) – The Generalized Equation for Organizational Direction, derived in Chapter 2.7, on Qualified Determinism.

Reproducing Equation 2.7.4:

$$Org\_Dir = DI \left(\begin{bmatrix} M_3 \rightarrow System_X \\ (\uparrow F \rightarrow I) \\ M_2 \rightarrow S_{System_X} \\ (\uparrow Sig \rightarrow F) \\ M_1 \rightarrow Sig_x \\ (\uparrow > P_P) \\ U \rightarrow x_U \end{bmatrix} \bigg|_{x=p,v,m,i} \right) \rightarrow$$

$x\_matrix_{strongest} @ level_{strongest}$

The activation-state can be thought of as being invoked when $level_{strongest}$ is at least M_1. Hence, as in Equation 4.2.1:

148

$$Activation - State_{condition} \geq M_1$$

Eq 4.2.1: Condition for Activation-State

Note that the Time-Matrix, in the lower-left side of (2.8.3),

specifies that approximately, adaptability may tend to be more automatic up to the emergence of cellular organisms. This is specified by the direction $M_3 \rightarrow U$, which indicates that it is the meta-level that organizes activity at U. At the human level the direction is flipped as specified by $U \rightarrow M_3$, and suggests that will, intention, feeling and the like are more important in stimulating the process of adaptability.

The visceral urge, then, may stimulate interaction with a collective-intelligence or 'specific-species-intelligence meta-function' that allows the generation of a new and specific 'predator-overcoming meta-function' so that the hypothetical insect in question goes through an adaptation to survive at least some kinds of predator attacks. This may take the form of this species of insect creating a hard-shell around it. Mathematically this new meta-function may take the following form as suggested by Equation 4.2.2:

$$Sig_{hard-protective-shell} = Xa + \overline{Yb_{0-n}} \text{ where } \begin{bmatrix} X \in [S_{System_P}] \\ Y \in [S_{System_{Pr}}, S_{System_P}, S_{System_K}, S_{System_N}] \\ a, b \text{ are integers}; a > b \end{bmatrix}$$

Eq 4.2.2: Hard Protective Shell Meta-Function

Specifically, S_{System_P} relates to the set of Polysaccharides and suggests that the chains of sugar molecules will adapt to become a shell to protect the insect. The primary element X is therefore an element of the set or Polysaccharides. $S_{System_{Pr}}$ refers to the set of Proteins. S_{System_K} refers to the set of Nucleic Acids. S_{System_N} refers to the set of Lipids. Y as the secondary element will invoke the action of some proteins, some existing polysaccharides, some nucleic acids, and some lipids in bringing about the adaptation as specified by $Sig_{hard-protective-shell}$. Perhaps the nucleic acids will code or coordinate how the proteins will work with the existing polysaccharides and lipids to create a new arrangement of polysaccharides that becomes the protective layer for the insect. This new arrangement of polysaccharides then becomes the purpose of a particular new type of specialized cell that exists to protect the insect against certain types of predators.

So while the breaking of patterns as specified by the Space-Matrix was a first step in the process, the subsequent creation of a new meta-function can be thought of as a second step. This new meta-function can be thought of as happening due to some combined functioning of the Light and Space Matrices. Specifically in the Space-Matrix, the breaking of patterns, $(\uparrow > P_x)$, allows the dynamics of M_1 to become active: $M_1 \rightarrow Sig_x$. This allows the ever-present layers of light, and specifically $C_{N:}\ f(S_{Pr} \times S_{Po} \times S_K \times S_H)$ in the Light-Matrix to become active so that new meta-function is generated.

The third step is suggested by ($\downarrow R_{C_U} = R_{C_N}$) that allows specific quantization to occur. It is proposed that the hard-protective-shell adaptability becomes real through a series of tightly coordinated space, time, energy, and gravity quantization as specified by (3.6.2-5) derived in Chapter 3.6. These equations are reproduced here for convenience and suggest the mechanics for the adaptability along this specific line of development in the larger scheme of Natural History.

Equation 3.6.2 models space-quantization:

$$Space_{quantization} = h_{UK}(Xa + \overline{Yb_{0-n}})$$

where: $\begin{bmatrix} X \in [S_{System_K}] \\ Y \in [S_{System_{Pr}}, S_{System_P}, S_{System_K}, S_{System_N}] \\ a, b \text{ are integers}; a > b \end{bmatrix}$

Recall that space-quantization further structures space so that seeds of knowledge or knowledge-potential to do with the new meta-function are now resident in the material ecosystem associated with the meta-function. In such a manner the species that set this meta-function into action will more and more tap into the possibilities to bring about the new protective mechanism it seeks.

Equation 3.6.3 models time-quantization:

$$Time_{quantization} = h_{UP}(Xa + \overline{Yb_{0-n}})$$

$$where: \begin{bmatrix} X \in [S_{System_P}] \\ Y \in [S_{System_{Pr}}, S_{System_P}, S_{System_K}, S_{System_N}] \\ a, b \text{ are integers}; a > b \end{bmatrix}$$

Recall that time-quantization alters the structure of time so that phases of maturity associated with seeds in space are now encoded in time. Slow and faster periods of change are in this way connected to the species seeking to protect itself against predators.

Equation 3.6.4 models energy-quantization:

$$Energy_{quantization} = h_{UPr}(Xa + \overline{Yb_{0-n}})$$

$$where: \begin{bmatrix} X \in [S_{System_{Pr}}] \\ Y \in [S_{System_{Pr}}, S_{System_P}, S_{System_K}, S_{System_N}] \\ a, b \text{ are integers}; a > b \end{bmatrix}$$

Recall that energy-quantization allows matter to be formed, and in this case, will have to do with the incremental material changes that the species will go through in forming a hard protective shell.

Equation 3.6.5 models gravity-quantization:

$$Gravity_{quantization} = h_{UN}(Xa + \overline{Yb_{0-n}})$$

$$where: \begin{bmatrix} X \in [S_{System_N}] \\ Y \in [S_{System_{Pr}}, S_{System_P}, S_{System_K}, S_{System_N}] \\ a, b \text{ are integers}; a > b \end{bmatrix}$$

Recall that gravity-quantization has to do with inter-

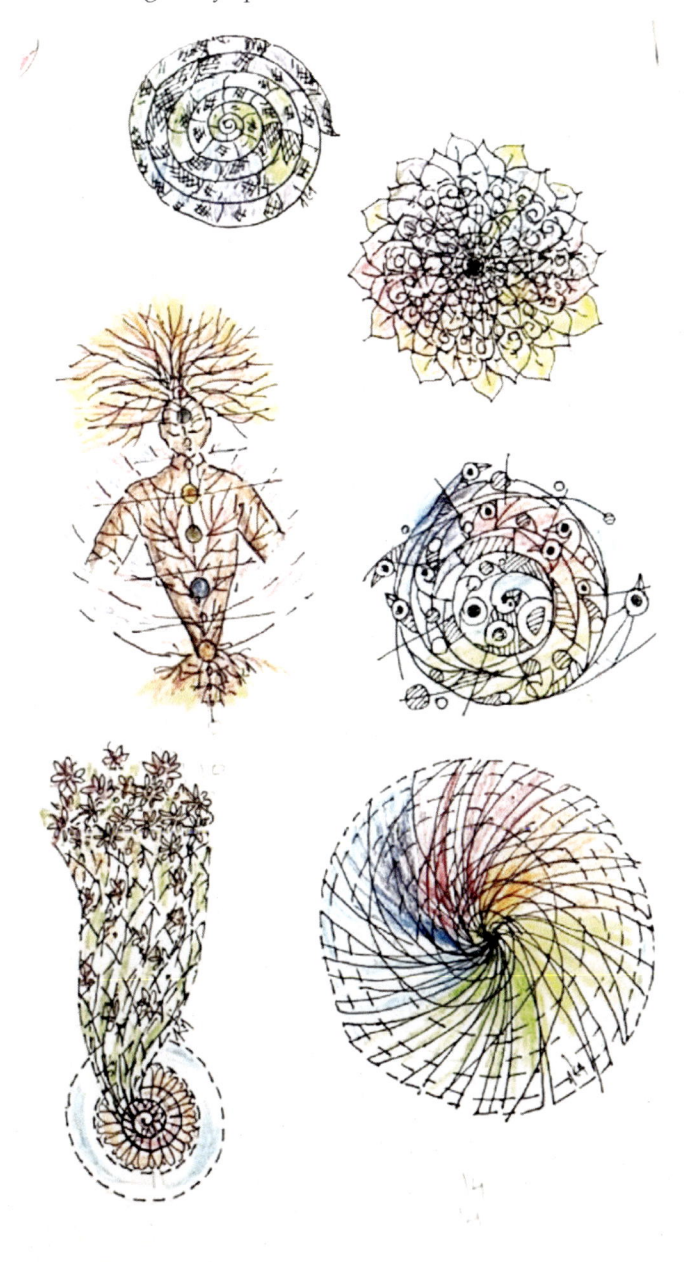

153

relation between the species and surrounding objects and will change the very nature of gravity to allow a subtle new balance in the species interaction with its surrounding so that the deep urge of the species can more easily be fulfilled.

Such quantization of space, time, energy, and gravity, describes the state of quantum-certainty. Quantum-certainty is the result of an intimate interaction with Q, the worlds accessed through quanta. The process of quantization and the state of quantum-certainty initiated through an activation-state, and the consequent enhancement or creation of a new meta-function, give a more intimate sense of the tightly coherent system we live in. Urge or desire, if deep enough, can set into motion an inter-relation between easily distinct layers of existence, allowing a deeper 'system-being' to intervene in regular system functioning.

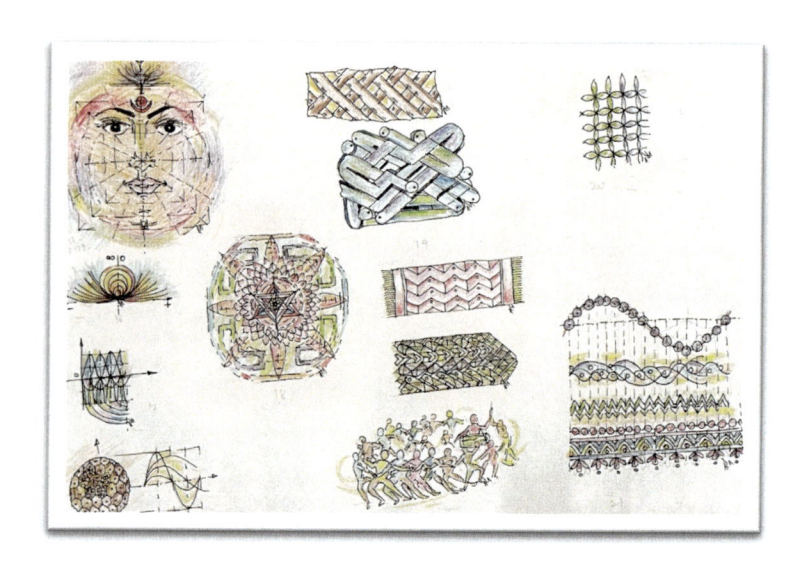

SECTION 5: QUANTUM-CERTAINTY & SUSTAINABLE HUMAN HISTORY

This section explores the process of quantum-certainty in determining sustainable human history.

First in Chapter 5.1, some key aspects responsible for a person's or a collectivity's action will be explored through the dynamics of thoughts, urges, feelings, and sensations.

Then Chapter 5.2 will explore signatures and uniqueness of countries.

Chapter 5.3 will illustrate, by an example leveraging the concepts and equations in the previous chapters, the process of quantum-certainty in determining how sustainable human history may play out.

Chapter 5.1: Sensations, Urges, Feelings, Thoughts

As human beings we experience sensations, urges and desires and wills, feelings and emotions, and thought. These are key aspects of our being and becoming and critical aspects of how choice at both the individual and collective levels may be determined.

This chapter will go over these fundamental aspects of self.

Sensations

Sensations are those things we experience with our senses. We see things, hear things, and smell things, taste things, can touch things. This ability to enter into relationship with objects through sensation is nothing other than a result of the emergence of Light's property of Presence. We become present to Presence through the device of sensation. Sensation can be thought of as the means by which this property of Light – Presence - molds or ingrains itself in us as human beings. Its potentiality, all which is contained in this aspect of Light, becomes available to us through the power of sensation. Hence an equation, Equation 5.1.1, will generally represent the family of sensations. Some elements that it would comprise of may be 'tangible', 'take notice of', amongst others.

$$Sig_{sensation} = Xa + \overline{Yb_{0-n}} \quad \text{where} \quad \begin{bmatrix} X \in [S_{System_{Pr}}] \\ Y \in [S_{System_{Pr}}, S_{System_P}, S_{System_K}, S_{System_N}] \\ a, b \text{ are integers}; a > b \end{bmatrix}$$

Eq 5.1.1: Sensation

There could also be equations for hearing, seeing, tasting, touching, and smelling.

But there is also a deeper experience of sensation that is possible. When we see things, for instance, what are we seeing? Is it just the surface rendering of the play of matter, or do we see that the fullness of Light is still there, with all its potentiality and possibility, in the smallest thing we look at? Do we see that the whole universe and more is present in all its fullness in the least thing that we easily ignore, or belittle, or loathe? When we touch things is it the seeming concreteness of the play of the particles or atoms or chains of molecules that we touch? Or is it the Love and Light and the vastness of all that IS that allows itself to be as a small corner that we touch so as to make infinity be felt by something so finite?

Such a deeper contact offered through sensation suggests a subset of (5.1.1) with secondary elements perhaps described as 'fullness of Light', 'contacting infinity', amongst others, thus also yielding an equation form (5.1.2):

$$Sig_{deeper-sensation} = Xa + Yb_{0-n} \text{ where } \begin{bmatrix} X \in [S_{System_{Pr}}] \\ Y \in [S_{System_{Pr}}, S_{System_P}, S_{System_K}, S_{System_N}] \\ a, b \text{ are integers}; a > b \end{bmatrix}$$

Eq 5.1.2: Deeper Sensation

Urges, Desires & Wills

Urges and desires and wills are similarly a play of the emergence of Light's property of Power. In the mystery of focus, the vastness of Light has projected itself in us into an apparent smallness that is in reality everything that is. And this smallness is trying through urge and desire and will to connect viscerally or even intentionally to other smallnesses that similarly are nothing other than the fullness of Light projected into a small smorgasbord of selected function. So the urge or desire for food, or companionship, or of possession, or of climbing a peak, is nothing other than Light's compressed property of Power, trying to reach more of the fullness that it is through a fulfillment of the urge or desire or will that it masquerades as. Hence urges can be represented as Equation 5.1.3:

$$Sig_{urges} = Xa + \overline{Yb_{0-n}} \quad where \begin{bmatrix} X \in [S_{System_P}] \\ Y \in [S_{System_{Pr}}, S_{System_P}, S_{System_K}, S_{System_N}] \\ a, b \text{ are integers}; a > b \end{bmatrix}$$

Eq 5.1.3: Urges

Elements may be of the type of 'grasp', 'possess', 'deeply connect', amongst others.

Feelings & Emotions

Feelings and emotions are a play of the emergence of Light's property of Harmony or Nurturing. Its instrument is the Heart and it generates an array of emotions that are an indication or active radar of whether we are moving toward or away from a reality of harmony, whether based on our small self or some larger Self of Light. Gradually, by navigating with these emotions and feelings we can get to a state where we always feel positive emotions which basically means we have more truly entered into relationship with some larger continent of Light. An equation for feelings is as represented by Equation 5.1.4:

$$Sig_{feelings} = Xa + \overline{Yb_{0-n}} \text{ where } \begin{bmatrix} X \in [S_{System_N}] \\ Y \in [S_{System_{Pr}}, S_{System_P}, S_{System_K}, S_{System_N}] \\ a, b \text{ are integers}; a > b \end{bmatrix}$$

Eq 5.1.4: Feelings

Thoughts

Thoughts are a play of the emergence of Light's property of Knowledge. Through the thought we can become greater or conceptualize things greater or begun to enter into relationship with some things other than our small self. Thought allows us to connect to more "othernesses" or even the oneness of the reality of Light.

$$Sig_{thoughts} = Xa + \overline{Yb_{0-n}} \text{ where } \begin{bmatrix} X \in [S_{System_K}] \\ Y \in [S_{System_{Pr}}, S_{System_P}, S_{System_K}, S_{System_N}] \\ a, b \text{ are integers}; a > b \end{bmatrix}$$

Eq 5.1.4: Thoughts

So we can see that the very substance of our becoming – sensations, urges, wills, desires, feelings, emotions, thoughts – are nothing other than an emergence of the four properties of light of presence, power, knowledge, and harmony.

Chapter 5.2: Basis for a Sustainable Global Civilization

A brief look at history will reinforce the idea that it is typically maturity along multiple and distinct dimensions, represented by the four properties of light, and their rich interaction that allows civilizations to endure.

Thus, those civilizations that have endured typically have a balance of all four families (Sri Aurobindo, 1971). Civilizations that have become extinct typically have had a focus on few drivers of innovation. Jared Diamond proposes five interconnected causes of collapse that may reinforce each other: non-sustainable exploitation of resources, climate changes, diminishing support from friendly societies, hostile neighbors, and inappropriate attitudes for change (Diamond, 2005). But these five sources may also be thought of as symptoms that arise due to the failure to adopt the catholicity of the sources of innovation emanating from each of the four sets of families. Further, the historian Toynbee suggested that societies decay because of their over-reliance on structures that helped them solve old problems (Toynbee, 1961). It can be interpreted that being thus biased they are unable to adopt the catholicity of the sources of innovation emanating from each of the four sets of families.

161

Approaching Civilization from a big-picture, global basis though, the sustainability of humankind will be ensured by a balance of development amongst the four sets of forces. This means that countries must be unique and in such a way that their primary emergence is distributed amongst all four sets of forces. Further, and based on this uniqueness, there must be an open and healthy interaction amongst these centers of uniqueness.

As an illustration, the uniqueness of representative countries in equation form may be as follows.

India, for example, in its essence, may be thought of as having an exceptional capacity for penetrating behind the surface, and further of meaningfully synthesizing many streams of development. Its primary power may thus be

162

from the family of knowledge, with a strong secondary driver being its ability to create living harmonies. The equation for India will then likely be represented by Equation 5.2.1:

$$Sig_{India} = Xa + \overline{Yb_{0-n}} \text{ where } \begin{bmatrix} X \in [S_{System_K}] \\ Y \in [S_{System_{Pr}}, S_{System_P}, S_{System_K}, S_{System_N}] \\ a, b \text{ are integers}; a > b \end{bmatrix}$$

Eq 5.2.1: India (Knowledge Family)

Japan, in its essence, may be thought of as having a strong and noble warrior nature, along with a refined sense of aesthetics, amongst other qualities. Its equation, Equation 5.2.2 would be of the form:

$$Sig_{Japon} = Xa + \overline{Yb_{0-n}} \text{ where } \begin{bmatrix} X \in [S_{System_P}] \\ Y \in [S_{System_{Pr}}, S_{System_P}, S_{System_K}, S_{System_N}] \\ a, b \text{ are integers}; a > b \end{bmatrix}$$

Eq 5.2.2: Japan (Power Family)

UK, in its essence, may be thought of as having a strong ability to create practical, materialistic harmonies, resulting in such things as working parliaments and advanced democracy, for example. Its equation, Equation 5.2.3, would be of the form:

$$Sig_{UK} = Xa + \overline{Yb_{0-n}} \text{ where } \begin{bmatrix} X \in [S_{System_N}] \\ Y \in [S_{System_{Pr}}, S_{System_P}, S_{System_K}, S_{System_N}] \\ a, b \text{ are integers}; a > b \end{bmatrix}$$

Eq 5.2.3: UK (Nurturing Family)

Thailand, in its essence, may be characterized by an exceptional sense of hospitality and sweet service, with an attention to detail in the practical arrangement of things. Its equation, Equation 6.3.4, would be of the form:

$$Sig_{Thailand} = \overline{\frac{Xa+}{Yb_{0-n}}} \text{ where } \begin{bmatrix} X \in [S_{System_N}] \\ Y \in [S_{System_{Pr}}, S_{System_P}, S_{System_K}, S_{System_N}] \\ a, b \text{ are integers}; a > b \end{bmatrix}$$

Eq 5.2.4: *Thailand (Service Family)*

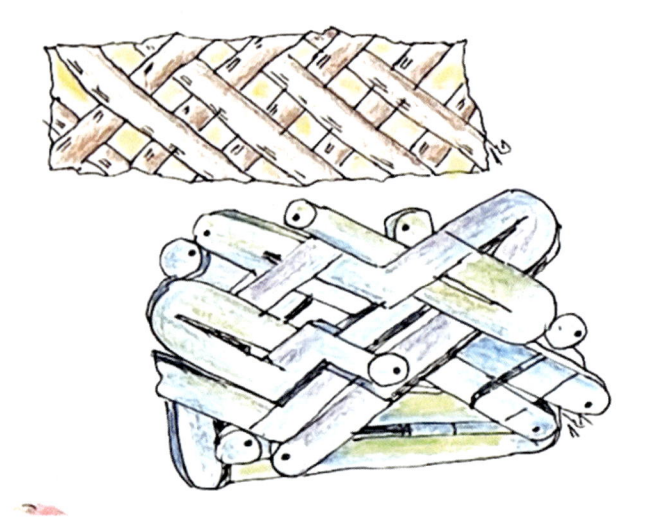

Similarly every nation on earth will have a uniqueness that can be represented by equations belonging to one of the four families.

The second requirement for a sustainable global civilization is a rich and mature interaction between unique nations. But as discussed in Chapter 2.5 on Emergence of Uniqueness, for uniqueness to emerge and mature is a process. The process is represented by Equation 2.5.1 reproduced here for convenience:

164

$$Sig_E = X \begin{vmatrix} C: Sig * mod\left(\int = 1\right) \\ F: Sig\ mod\ (c) \\ I: Sig\ mod\left(\int \overline{G, e, \pi}\right) \\ M: Sig * mod\ (G) \\ V: Sig * mod\ (e) \\ P: Sig * mod\ (\pi) \end{vmatrix}$$

Equations 5.2.1 – 4 represent examples of the essence of uniqueness. For these to become living practicalities requires work at many different levels within a nation. If a nation has not really done the work to animate itself with its uniqueness then it may exist in the physical or P-state, or vital of V-state, or mental or M-state, which by definition lack sufficient maturity, and interaction with other nations is then going to be compromised. Moving to the integral or I-state, will allow a nation to at least not be locked in to a point of view. A consistent I-state practiced by each nation then is the minimum requirement for a sustainable global civilization.

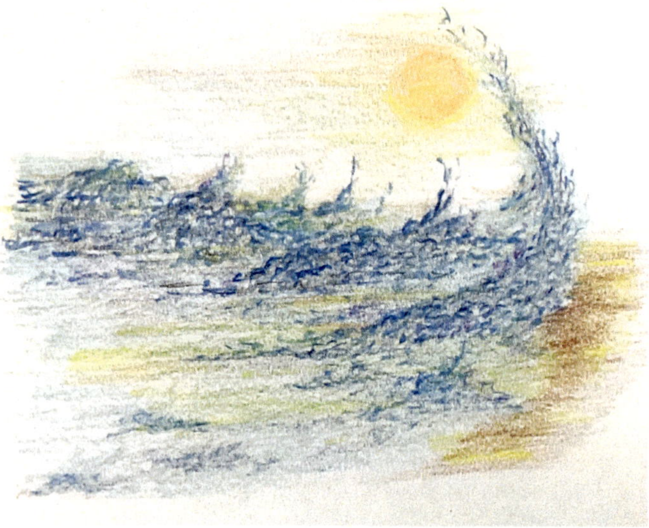

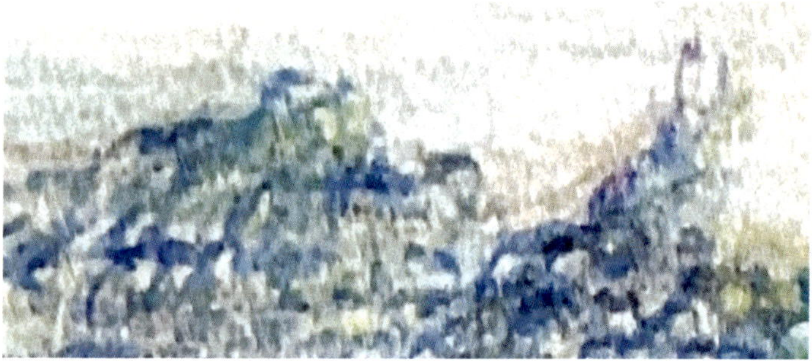

Chapter 5.3: Quantum-Certainty & Sustainable Human History

The fundamental question of a contemporary human history in the making is whether our global civilization will become sustainable. Additional insight into process and mechanics for such an endeavor can perhaps be gained by leveraging some of the mathematical tools developed through this book.

Natural History is by definition sustainable because the being and becoming of plants and animals is coordinated to a much higher degree by Nature herself. In humans, Nature has separated the will and other ego-based faculties from her intricate and automatic workings, and human choice begins to become far more important. Human history therefore never has to be sustainable. In fact it is a miracle that it is. But the mathematics in this book seeks to describe the workings of the 'miracle'. There is a state of quantum-certainty that has to result in order for human history to become sustainable. The possibility of quantum-certainty is initiated with an activation-state. The activation-state will open a quantum-window to allow interaction with the worlds of meta-function behind the surface layer U. It is the inter-relation with such meta-functions that creates and will continue to create sustainable human history. These meta-functions organize their own reality through a detailed process of multi-dimensional quantization that will in effect change the nature of experienced reality.

This chapter will explore the following key concepts:
1. Necessity of quantum-certainty in making intended history, leveraging off the framework applied in Chapter 4.2.
2. Use of the qualified determinism framework developed in Chapter 2.7 to understand the current state of global affairs.
3. Dissection of some dynamics of moving nations to an I-state, as explored in Chapter 5.2.

The Light-Space-Time Emergence equation (2.8.3) derived in Chapter 2.8, and reproduced here for convenience, will provide some insight into the dynamics of the pregnant-infinity, and of quantum-certainty in the making of history. Reproducing the equation:

$$Emergence_{light-space-time} =$$

$$\begin{bmatrix} \begin{bmatrix} C_\infty: [Pr, Po, K, H] \\ (\downarrow R_{C_K} = R_{C_\infty}) \\ C_K: [S_{Pr}, S_{Po}, S_K, S_H] \\ (\downarrow R_{C_N} = R_{C_K}) \\ C_{N:} f(S_{Pr} \times S_{Po} \times S_K \times S_H) \\ (\downarrow R_{C_U} = R_{C_N}) \\ C_U: [\leftarrow, |, \rightarrow, \leftrightarrow] \end{bmatrix}_{Light} \begin{bmatrix} M_3 \to System_X \\ (\uparrow F \to I) \\ M_2 \to S_{System_X} \\ (\uparrow Sig \to F) \\ M_1 \to Sig_x \\ (\uparrow > P_x) \\ U \to x_U \end{bmatrix}_{Space} \\ \begin{bmatrix} M_3 : All\ t \\ \downarrow \\ M_2 : 0 \geq t > \infty \\ \downarrow \\ M_1 : 0 > t > \infty \\ \downarrow \\ U \to \begin{array}{l} t \sim E_{Cell}; TC: M_3 \to U \\ t \sim E_{Human}; TC: U \to M_3 \end{array} \end{bmatrix}_{Time} \quad TC \to x_T \end{bmatrix} \langle x_U | x_T \rangle$$

The Light-Matrix, the top-left matrix in (2.8.3), suggests the possibilities in the pregnant-infinity. These possibilities, recall, have been set up by realities so created by light traveling at different speeds. While

169

prevalent dynamics at the surface layer, U, are always influenced by dynamics from the deeper layers, there is an activation-state that can be created that will allow a more focused and intentional activation or creation of the meta-functions in the deeper layers.

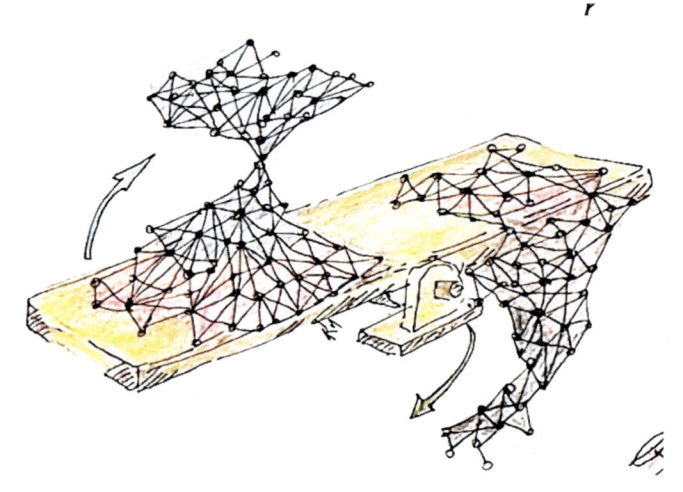

r

The Space-Matrix, the top-right matrix in (2.8.3), suggests the dynamics involved in creating such an activation-state. Essentially this involves the overcoming of patterns at U.

Leveraging (2.7.3) derived in Chapter 2.7, gives insight into understanding the prevalent patterns at the global level. This equation assesses the overall direction an organization will move in given the current set of forces it is subject to. The equation is reproduced below for convenience

$$Org\_Dir = DI \begin{pmatrix} \begin{bmatrix} M_3 \rightarrow System_{Pr} \\ (\uparrow F \rightarrow I) \\ M_2 \rightarrow S_{System_{Pr}} \\ (\uparrow Sig \rightarrow F) \\ M_1 \rightarrow Sig_P \\ (\uparrow > P_P) \\ U \rightarrow Physical_U \end{bmatrix} \begin{bmatrix} M_3 \rightarrow System_P \\ (\uparrow F \rightarrow I) \\ M_2 \rightarrow S_{System_P} \\ (\uparrow Sig \rightarrow F) \\ M_1 \rightarrow Sig_V \\ (\uparrow > P_V) \\ U \rightarrow Vital_U \end{bmatrix} \\ \begin{bmatrix} M_3 \rightarrow System_S \\ (\uparrow F \rightarrow I) \\ M_2 \rightarrow S_{System_S} \\ (\uparrow Sig \rightarrow F) \\ M_1 \rightarrow Sig_M \\ (\uparrow > P_M) \\ U \rightarrow Mental_U \end{bmatrix} \begin{bmatrix} M_3 \rightarrow System_N \\ (\uparrow F \rightarrow I) \\ M_2 \rightarrow S_{System_N} \\ (\uparrow Sig \rightarrow F) \\ M_1 \rightarrow Sig_I \\ (\uparrow > P_I) \\ U \rightarrow Integral_U \end{bmatrix} \end{pmatrix} \rightarrow$$

$x\_matrix_{strongest}$ @ $level_{strongest}$

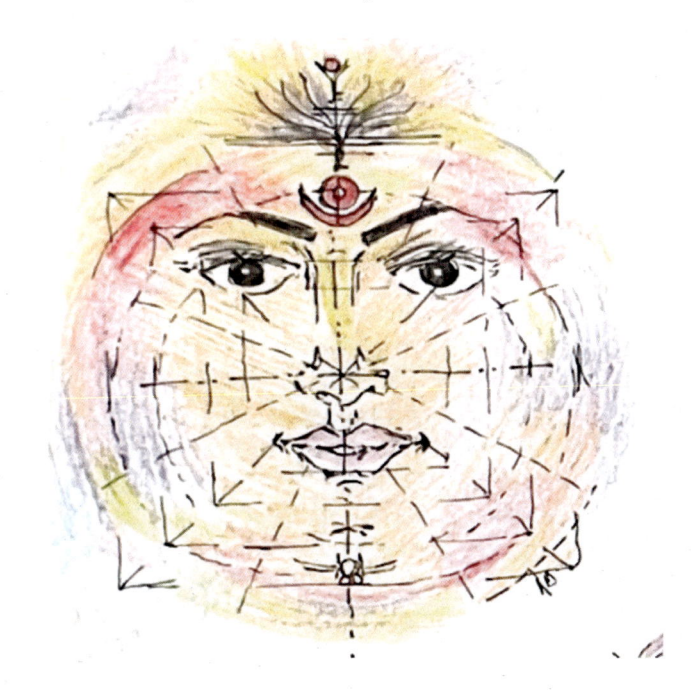

The equation comprises four matrices, each one dedicated to fundamental realities vying for domination. Hence, the top left-hand matrix, the Physical-Matrix, can be thought of as existing to keep the entities within the system in a reality of isolated separation. The top right-hand matrix, the Vital-Matrix, can be thought of as promoting interaction though anchored in separateness, that therefore could result in domination by any one entity. The bottom left-hand matrix, the Mental-Matrix, can be thought of as promoting some mental ideal. The bottom right-hand matrix, the Integral-Matrix, can be thought of as integrating the possibilities represented by the other three matrices together.

Within each matrix the operational level can be one of four, though usually it is at the first or U level. Within U sub-levels are likely - P, or V, or M - or the other levels as suggested by (2.5.1).

In a summary assessment, while each of the matrices or motive forces is globally active, it is likely the Physical-Matrix with the strong impetus for Nationalism bordering on fundamentalism that seems to be most active in today's environment. Hence today's global reality seems to be strongly animated by *Physical@U*.

This represents a global pattern that must be broken for the initial activation-state to result. Note that by definition such an activation-state necessitates moving in the direction where the speed of light increases. Note that speed of light increasing implies that the sense of separation is diminishing, and entities can have a more complete access to the fullness that they are. This can only happen when patterns get elevated. If patterns are in the U realm, being therefore untransformed, that will by definition not allow a quantum-window to open.

As per discussion in Chapter 4.2, this activation-state can be further specified by leveraging (2.7.4) – The

Generalized Equation for Organizational Direction, derived in Chapter 2.7, on Qualified Determinism.

Reproducing Equation 2.7.4:

$$Org\_Dir = DI \left(\begin{bmatrix} M_3 \rightarrow System_X \\ (\uparrow F \rightarrow I) \\ M_2 \rightarrow S_{System_X} \\ (\uparrow Sig \rightarrow F) \\ M_1 \rightarrow Sig_x \\ (\uparrow > P_P) \\ U \rightarrow x_U \end{bmatrix} \Bigg|^{x=p,v,m,i} \right) \rightarrow$$

$x\_matrix_{strongest}$ @ $level_{strongest}$

Activation-state can be thought of as being invoked when $level_{strongest}$ is at least M_1. Hence, as suggested by the previously derived (4.2.1), reproduced here for convenience:

$Activation - state_{condition} \geq M_1$

Further, and as per the discussion in the previous chapter, assume that there is a deep wanting felt by a threshold number of people. Perhaps this wanting, a step towards a global sustainable civilization, is for each nation to operate more from their true essence, or I-state, as opposed to fundamentalist form. In other words a meta-function, $Sig_{Nation-essence}$, must be created that will then organize its own reality, and as indicated by Equation 5.3.1.

$$Sig_{Nation-essence} = Xa + \overline{Yb_{0-n}}$$

$$where \begin{bmatrix} X \in [S_{System_K}] \\ Y \in [S_{System_{Pr}}, S_{System_P}, S_{System_K}, S_{System_N}] \\ a, b \text{ are integers}; a > b \end{bmatrix}$$

Eq 5.3.1: Nation-Essence Meta-Function

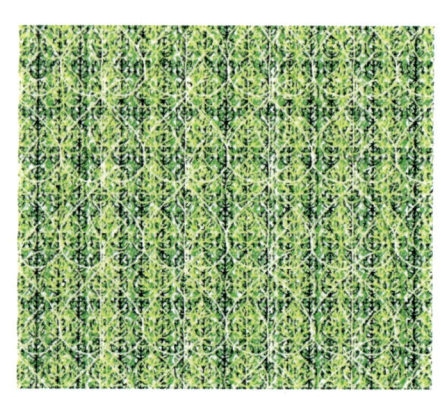

While the key aspect, or primary X-element in the nation-essence meta-function may be an element from the set of Knowledge, and specifically 'what the nation-essence means', there will be many parts, represented by the Y-elements, such as the structure, processes, institutions, culture, that will also need to be created. An important denominator in breaking patterns in each of these areas though is hinted at in Chapter 5.1, and involves the creation of 'cohesion' at the individual and collective levels as in Equation 5.3.2, Nation-Essence-Cohesion with the primary element belonging to the set of Harmony or Nurturing:

$$Sig_{Nation-essence-cohesion} = Xa + \overline{Yb_{0-n}}$$

$$where \begin{bmatrix} X \in [S_{System_N}] \\ Y \in [S_{System_{Pr}}, S_{System_P}, S_{System_K}, S_{System_N}] \\ a, b \text{ are integers}; a > b \end{bmatrix}$$

Eq 5.3.2: Nation-Essence-Cohesion Meta-Function

Hence, S_{System_K} relates to the set of Thoughts required to contextualize and frame nation-essence. $S_{System_{Pr}}$ refers to the set of Sensations required to assess and interpret the constantly arising signs of the new development – the sensory cues as it were that will allow any individual or collectivity to sense that they were on the right path. S_{System_P} refers to the set of Urges and Wills constantly required to ensure that the goal of cohesion was attained. S_{System_N} refers to the set of Feelings and Emotions that will need to be generated to attain cohesion.

As such cohesion becomes a reality, older patterns will break down and there will be easier and longer activation-state periods, potentially allowing 'miracle' to become the law.

As suggested in Chapter 4.2, while the breaking of patterns as specified by the Space-Matrix was a first step in the process, the subsequent creation of a new meta-function can be thought of as a second step. This new meta-function can be thought of as happening due to some combined functioning of the Light and Space Matrices. Specifically in the Space-Matrix, the breaking of patterns, $(\uparrow > P_x)$, allows the dynamics of M_1 to become active: $M_1 \rightarrow Sig_x$. This allows the ever-present layers of light, and specifically $C_{N:} f(S_{Pr} \times S_{Po} \times S_K \times S_H)$ in the Light-Matrix to become active so that new meta-function is generated.

The third step is suggested by (↓ $R_{C_U} = R_{C_N}$) that allows specific quantization to occur. Quantization is of fundamental importance because it is what allows the fabric of experienced reality to change. The new thought to do with nation-essence is quantized as seeds of knowledge in space, fundamentally changing the structure of space. Hence, reproducing Equation 3.6.2, space-quantization is modeled as:

$$Space_{quantization} = h_{UK}(Xa + \overline{Yb_{0-n}})$$

where: $\begin{bmatrix} X \in [S_{System_K}] \\ Y \in [S_{System_{Pr}}, S_{System_P}, S_{System_K}, S_{System_N}] \\ a, b \text{ are integers}; a > b \end{bmatrix}$

The phases of maturity that the seeds of knowledge will go through give time a new reality or structure. In other words time too is re-oriented to promote the outcome of the new meta-function in effect. This quantization of time, essentially giving time in this new fabric a different meaning is modeled by Equation 3.6.3, reproduced here for convenience:

$$Time_{quantization} = h_{UP}(Xa + \overline{Yb_{0-n}})$$

$$\text{where:} \begin{bmatrix} X \in [S_{System_P}] \\ Y \in [S_{System_{Pr}}, S_{System_P}, S_{System_K}, S_{System_N}] \\ a, b \text{ are integers}; a > b \end{bmatrix}$$

The new meta-function changes the very nature of matter, which is quantized to allow the material fabric of existence to promote the new meta-function in unimaginable ways. This quantization of energy, resulting in a subtly different matter is modeled by Equation 3.6.4, reproduced here for convenience:

$$Energy_{quantization} = h_{UPr}(Xa + \overline{Yb_{0-n}})$$

$$\text{where:} \begin{bmatrix} X \in [S_{System_{Pr}}] \\ Y \in [S_{System_{Pr}}, S_{System_P}, S_{System_K}, S_{System_N}] \\ a, b \text{ are integers}; a > b \end{bmatrix}$$

As a result the very force of gravity is altered locally as well, so that its quantization promotes subtle interaction in the related ecosystem that will be predisposed to making the meta-function a reality. This quantization of gravity is modeled by Equation 3.6.5, reproduced here for convenience:

$$Gravity_{quantization} = h_{UN}(Xa + \overline{Yb_{0-n}})$$

$$\text{where:} \begin{bmatrix} X \in [S_{System_N}] \\ Y \in [S_{System_{Pr}}, S_{System_P}, S_{System_K}, S_{System_N}] \\ a, b \text{ are integers}; a > b \end{bmatrix}$$

It is the combined quantization that can be thought of as quantum-certainty. It is such quantum-certainty that will ensure sustainable human history.

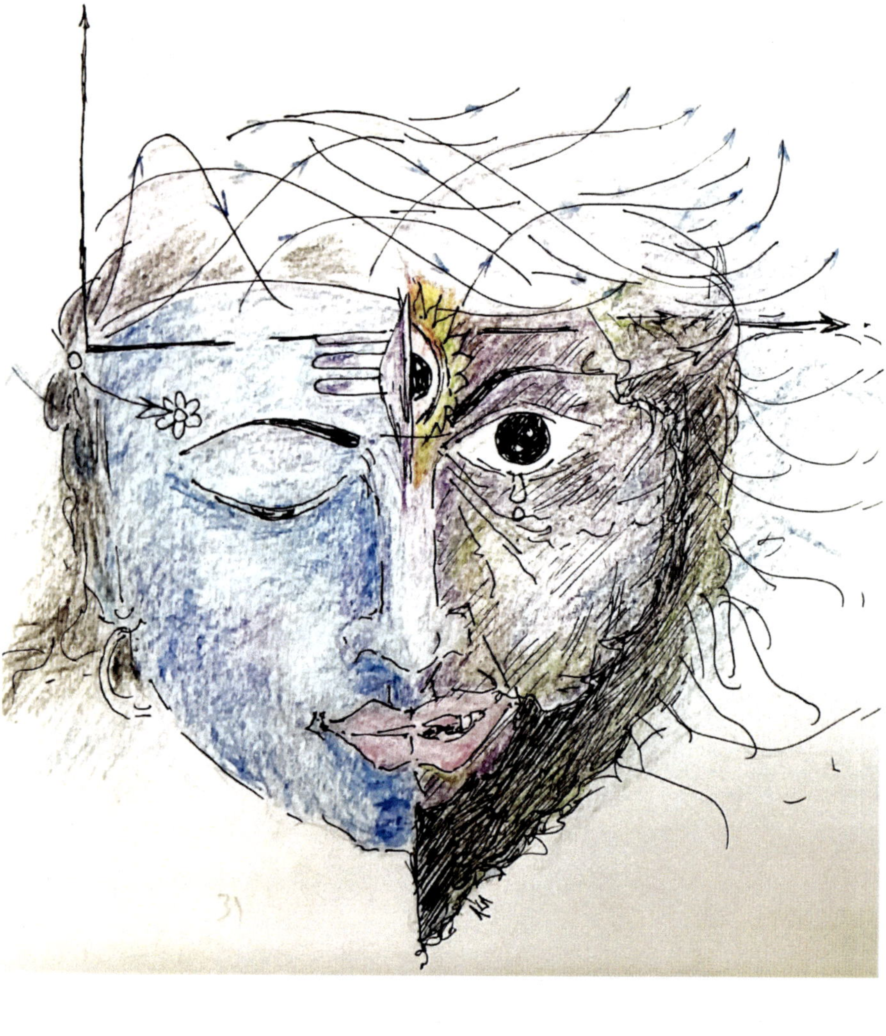

APPENDIX: A BRIEF EXPLANATION OF THE ILLUSTRATIONS IN THIS BOOK

(Narendra Joshi, Ph.D.)

Fig 1: Even if source is one there is no need to forcefully fit all things in one layer as that results in appearance of weirdness. Hence here many layers at different levels or speeds are drawn and they have interaction and connections with each other and with the uniting source which is also the center of their circular orbits.

Fig 2 : A multiple layer structure accessed through quanta creates quantum certainty. Creation of such a meta-function sets into motion a precise process of space, time, energy, and gravity quantization.

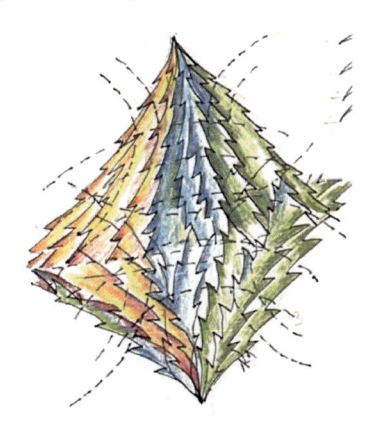

SECTION 1

Fig 3 and Fig 4: Different illustrations related to points upcoming in this section are put here in a collage.

Fig 5: Everything is made from Light - from our bodies, thoughts, to plants, animals, quarks - micro to macro. But how to visualize this as apparently everything looks so different? It can be seen as a hierarchy of continuous rising expressions of light. Four colors are used in most of the sketches in this part to express four components at each level forming reality, as is explained in this Cosmology of Light book-series and in this book as well.

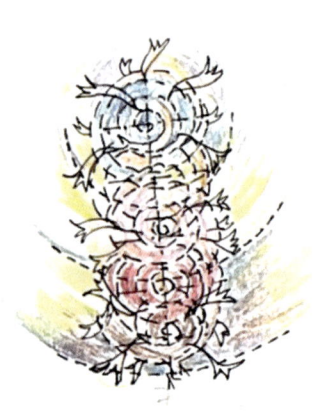

Fig 6: Light at 4 minutes from the Sun and at 8 minutes in its journey to the earth gives insight into different possibilities of creation.

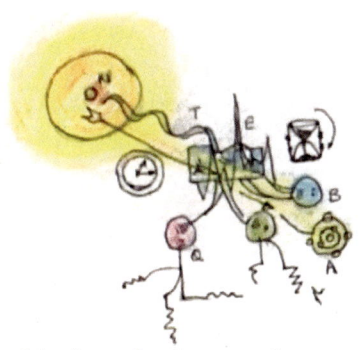

Fig 7: As Light is traveling at a finite speed and will take some distance or time to express what is in it, so there is a build-up of energy that forms packets or quanta - thereby the world of quarks, leptons, bosons, atoms is formed.

181

Fig 8: The Light at Infinity, if it is lit in a dark volume it instantly occupies every corner of it, removes darkness and gives life. Shown here is a circle of light with 4 'O's at four corners: Omniscient, Omnipotent, Omnipresence, Omninurturing.

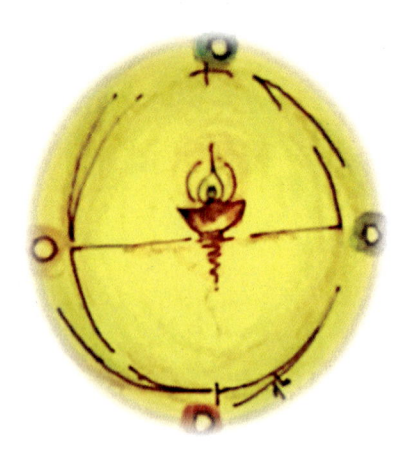

Fig 9 and Fig 10: Big bang is the slowing down of the speed of Light. There are experiments done to slow it down till it reaches a snail's pace. Shown in two figures here: a snail has a continuous spiral in her shell, which is also a sign of slowing down.

182

Fig 11 & 12: The four-fold reality - Presence, Power, Knowledge, Harmony; Physical, Vital, Mental, Integral; Past, Present, Future, Matter; Omni-presence, -science, -potent, -nurturing. See color scheme reiterated consistently.

Fig 13: Quanta is a doorway, a passage or interface for transformation, shown here as Light rays through a doorway often resembling ancient Indian temple dwar/praveshdwar (door/entrance).

183

Fig 14: Both Space and Time are not abstract but are highly structured and are in a pattern of four factors shown here in an efflorescence of different layers and elements.

Fig 15: Space consisting of a vast array of seeds derived from the properties of Light. Shown here is seed of a tree within which are many smaller seeds all interrelated with each other.

Fig 16: What appears as opposites actually are complimentary things, especially when they are turned from linear to circular or cyclical mode and seen from a higher point of view. Everything is in cycles and thus circular seed and its cyclical projections in other views represent the complementarity instead of opposition in Nature.

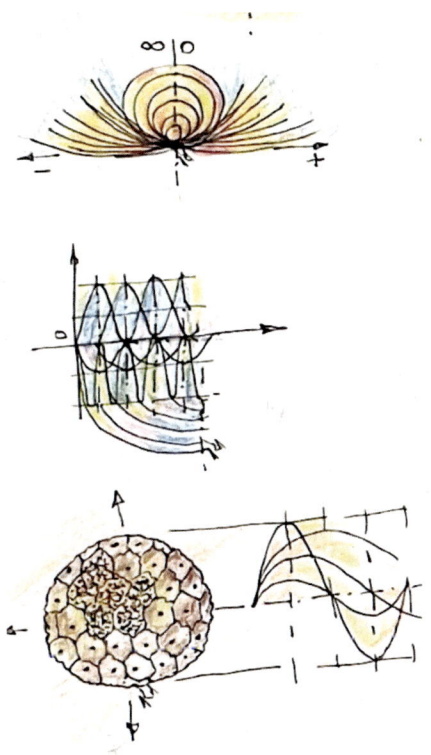

185

SECTION 2

Fig 17: Light Matrix is shown here as grilled window of rows and columns in which an assemblage of soil, seed, plant, flowerpot, flowers, butterfly and cat are seen enjoying Sun light with blue sky behind.

Fig 18 & 19: Connections are emphasized in this thought as they join the islands of matter: "for the reality of nature with a past, present, and future, to also be experienced as a phenomenon of connection between seemingly independent islands of matter.

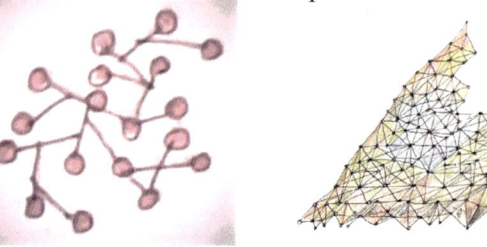

This characteristic of 'connection' is therefore also proposed to be implicit in the nature of light."

186

Fig 20 & 21: The four-fold reality and emerging outcome of it in similar four-folded ness is symbolized by waves of four layers and colors having similar patterns at all levels, and yet different properties. Also seen is a number of flowers emerging from a larger flower which in fact is also an expanding spiral. The emergence is in multiple curve paths which makes a matrix.

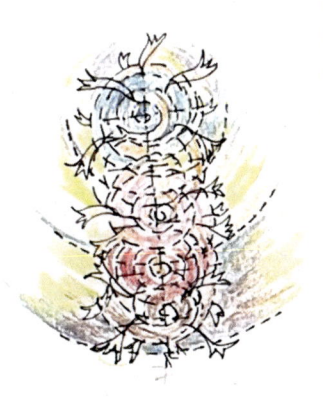

Fig 22: Illustration to show that Light at one point results instantaneously in light at other points

Fig 23: Connections used by joining bamboo or cane strips weaved in a shape of a basket or a dome is seen widely in arts of many folk communities of India. Such design is used to make cups, containers, headgears. Here one recollects that a pair of upper and lower hemispheres and an upside down utensil has a significance in Upanishadic/Vedantic images as well. This is also seen as images used in modern organizational design theories.

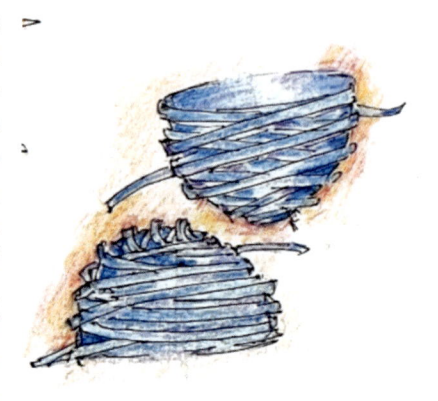

Fig 24: Light Matrix in equation 2.1.6 visualized as a 3D tower.

Read from top to bottom row as indicated by arrows.

188

Fig 25: Wave-Particle Duality turns into 'Quadrality' here with four components.

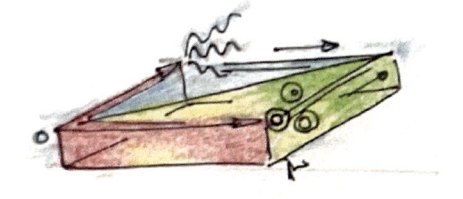

Fig 26: Visualizing Equation 2.2.1 - System Presence:

Physical leading to vital, leading to mental. Container of system (power, knowledge, harmony)

189

Fig 27: Equation 2.2.2 visualizing System Power

Fig 28: Equation 2.2.4 visualizing System Nurturing:

Fig 29: 'To be able to leverage or activate this fourfold intelligence at will is the ultimate act of innovation ': Four factors in four colors and forming four quadrants of a box are leveraged here to create here something innovative and out-of-the-box.

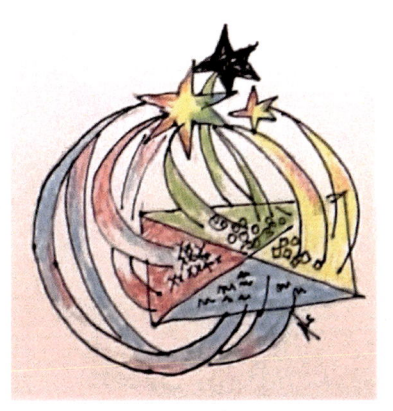

Fig 30: Uniqueness of an organization. A pattern like four colored mandala and petals used here for organic and harmonious self-expression. The downward and upward triangle intersecting at center symbolizes the soul aspiration from bottom and grace of Divine from top. Also gives a glimpse of the ancient mystic mantra: 'Oh Jewel in the Lotus.'

Fig 31: puts previous figure in another way of expression as often seen in yantra or in images used in meditation circles.

Fig 32: Emergence of fire flames from altar visualizing emergence of uniqueness in general.

192

Fig 33: Altars with carefully selected stones or bricks lay down the base. A stone path made in water also is an image often used in organizational design.

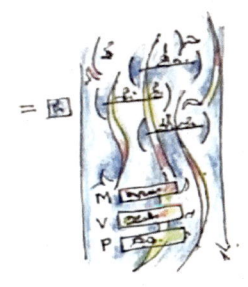

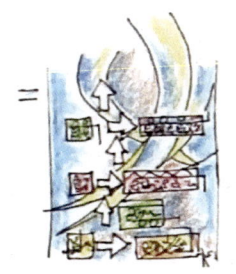

Fig 34: Equation 2.5.1 - 'Emergence of Uniqueness' is visualized here.

And the initial part of chapter 2.6 Inherent Dynamics of Any System

Equation 2.6.1 is visualized. The arrows from left to right and upward going in layers are seen here with suddenly rising curves.

193

Fig 35 & 36:

In visualizing adjacent levels that exist in paired relationships and inter-related functionality, and then turning it into a complete matrix.

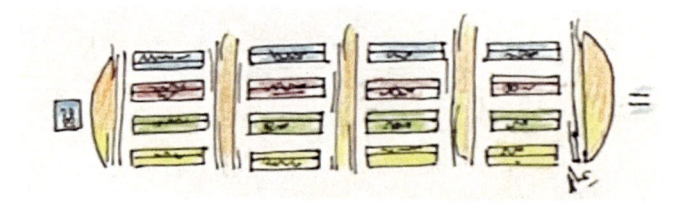

'A deeper dimension of a physical layer suggests synonymity with a meta-layer.'

Fig 37: Eq 2.6.2 - Transformation Circle represents innovation as visualized here.

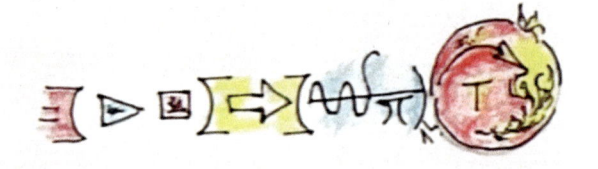

Fig 38: Equation 2.6.3/4/5 - Inherent Dynamics of Vital /Mental and Integral is visualized here.

Fig 39 & 40: Qualified determinism section is visualized with now putting together symbols and images used earlier, -such as circle and spiral, a spirally expanding space, a snail, a pair of upper and lower hemisphere, a sandclock, linearity turning into circularity and that into conical structure, etc. Metalayers of these different symbols, when integrated, results in some images which are often discussed in the Upanishads and Bhagvad Gita to describe how the existence is created from the Almighty.

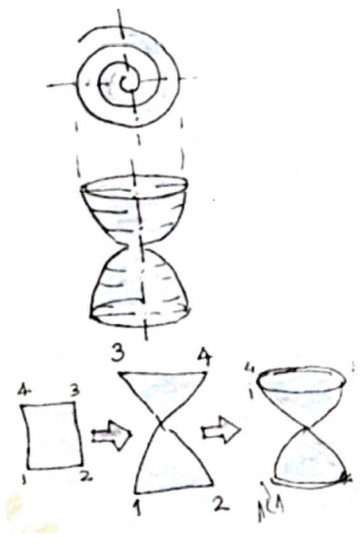

195

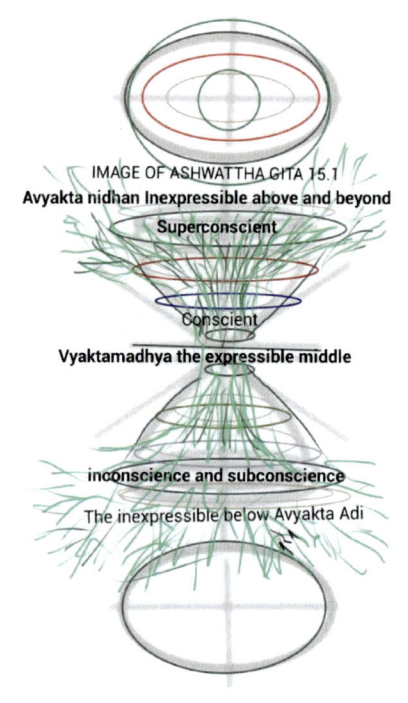

One such image most startlingly fitting is that of an Ashwattha tree. It is taken from the 15th chapter of the Gita (15.1 : first verse from chapter 15). Here Ashwattha, an inverted tree with its infinite roots above and infinite branches below represents the infinite diversity of this cosmic existence and infinite Unity whose projection it is. This diversity is ascending slowing on evolutionary patterns to reach the Unity. The middle of it is expressible while the upper and lower are Inexpressible…this also matches images of mandala and its elevation views in Book 1 of this Cosmology of Light book series.

IMAGE OF ASHWATTHA GITA 15.1
Avyakta nidhan Inexpressible above and beyond
Superconscient
Conscient
Vyaktamadhya the expressible middle
inconscience and subconscience
The inexpressible below Avyakta Adi

This point, as planned, will be further worked upon along with other images of the Upanishads and the Gita, and with quantum science and Sri Aurobindo's integral vision in coming parts.

Fig 41 & 42 are about organizational direction seen through the Indian architectural elements.

Going through Equations 2.7.1, 2.7.2 and 2.7.3 reaching then to equation 2.7.4 which is about the generalized equation for organizational direction, is visualized through the images resembling Indian ancient and medieval architecture and various stone carvings in it. These elements also have multilayers, gaps and pauses between crowding details, interrelated structures and so on.

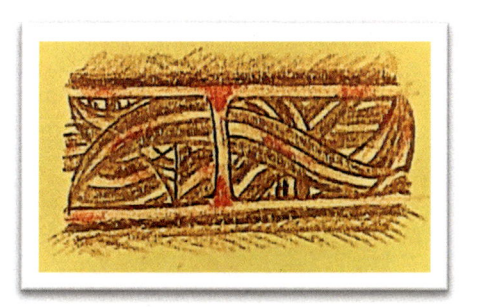

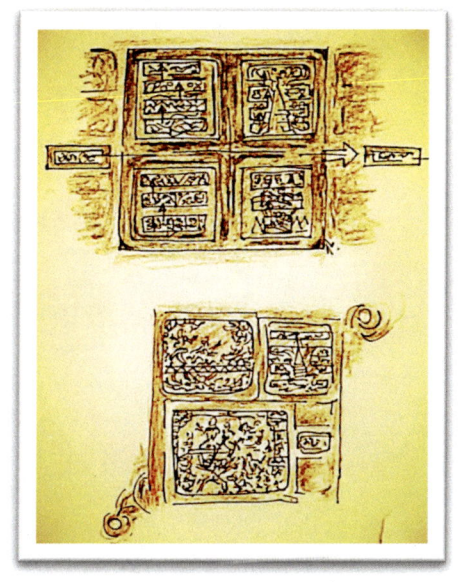

Elements of such Complex Adaptive Systems (CAS) are active, connected, interdependent, and emergent.

Fig 43: Based on this exploration we come to equation 2.8., that is, Generalized Equation of Innovation.

It is visualized again with ancient and medieval Indian architecture and iconography, Vastu Kala and Shilpkala. Especially breathtaking are miniaturized carvings and detailed depictions in it - so layered, intricate, and yet connected on different dimensions, resulting in wholeness. The blooming lotus pair shows the sheer bliss of innovation after such hard work.

SECTION 3

Figure 44, 45 & 46: Interpretation of Quantum Phenomena.

'You view the world through a model always…'

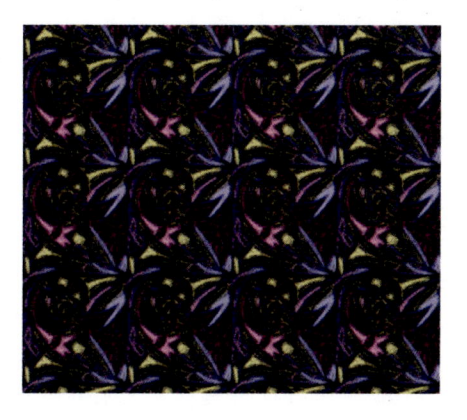

The whole existence is predetermined, predesigned, predestined emphasized Sri Aurobindo. It is qualified determinism. Intricate reason expressed mathematically ultimately resulting in delicate patterns of wisdom, strength, beauty, harmony, and perfection.

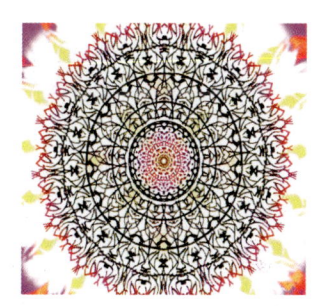

199

The linear row and column patterns in Fig 44 is further taken to four quadrant pattern in Fig 45 and is made integral in the mandala pattern in Fig 46.

Fig 47: Schrodinger's famous cat experiment is seen here with a novel way of a cat in a cage as cropped from the main image of the Light matrix seen earlier.

Fig 48: What was thought to be a thought experiment aimed at exposing absurdity of a quantum phenomenon in fact proved to be a classic way of explaining it. A seemingly chaotic design here in Fig 48 turns out to be well formulated one if seen more closely.

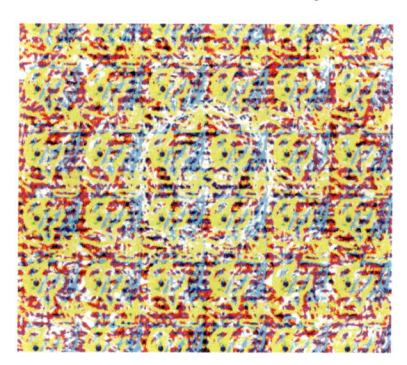

200

Fig 49: 'It is not hierarchy of parts; it is hierarchy of wholes.' 'The Universe thus appears as a complicated tissue of events where relations intersect and give birth to objects.' Visualized here with a three dimensional matrix of connections and ascending relations – with four colors used to remind us of four foldedness - and cross connections that emphasize that lateral, nonlogical, and nonlinear is the norm.

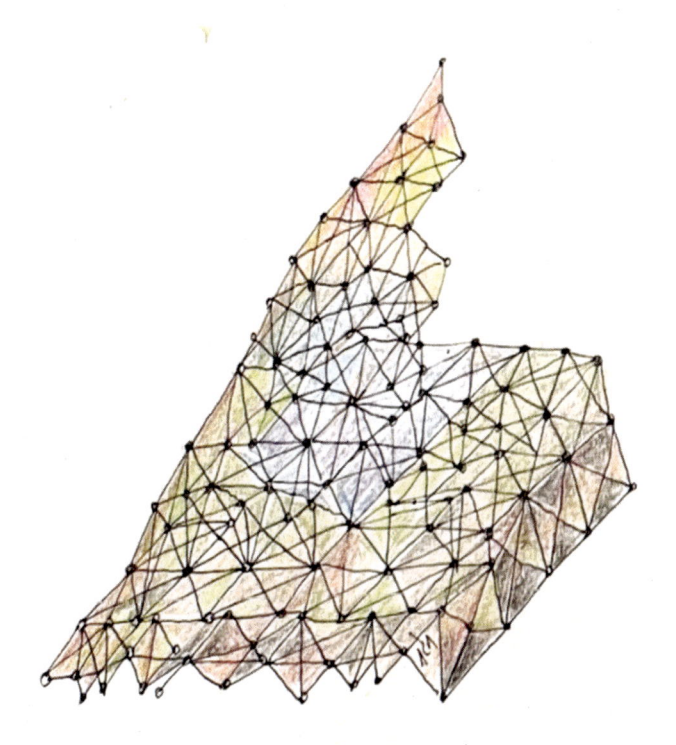

Fig 50: A collage of some of the classic representations from ancient and medieval Indian architecture is put here, as it resembles strikingly some of the key ideas discussed in the book - pattern which repeats and connects, fractals, curves and lines mixed together, space used effectively for depicting time, and so on.

Fig 51: Duality turns into Quadrality.

Fig 52: Quantum tunneling is visualized here as light passing through a tunnel in a mountain range.

203

Fig 53: Schrodinger's 'What is Life' is visualized with a pattern of layered ascent with leaves and lotus flowers in it. Order from order. But order not just from and for dry reason but as sap of life giving nourishing and heartening beauty, love, and energy.

Fig 54: A pattern that connects curved space and time - connections defining objects through intersection. All such points are put together here in a pattern where four lines are used and four colors which interact and intersect endlessly, resulting organically in a hemispherical shape as we proceed. This again resembles a weaving in fabric or a craft from bamboo or cane as seen in Indian arts.

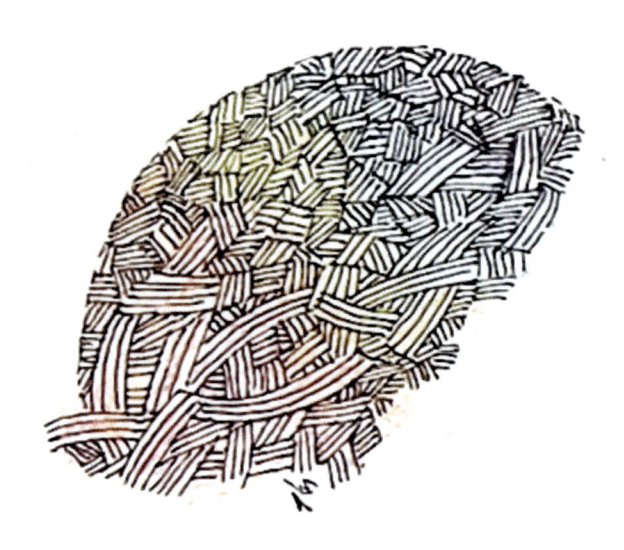

Fig 55: 'This quantum property has been invoked in describing how birds navigate across the earth ..' the typical curve traced by flock of birds shown here with red orange sky behind to depict the sunset time.

Fig 56: Birds flying back (in time) seen here with sun behind the grid formed by entanglement of the leaves of several coconut trees.

205

Fig 57: 'Admitting physics at each meta layer and knowing that all is One, Undivided, and Whole' is visualized here as 'Akhandamandalakar' (unbroken continuous expanding spiral) of ancient Indian texts.

Fig 58: Quanta as gateway or passage is multiplied here as quantizing multiple elements. Space, time, energy, gravity all in a linear pattern of multiple passages /openings as seen in architecture.

Fig 59: Light Matrix and Emergence is visualized as patterns originating from same pitch but resulting in startlingly diverse designs as we move from layer to layer.

207

Fig 60: 'The quantization-window, as it were, potentially allows the precipitation of, or inter-relation with, or creation of a cohesive and compelling meta-function or signature...' Here different layers are depicted as different waves on a seashore which are often seen as similar and

cohesive pattern, self-repeating in time and space. It is cyclic in its forward and backward movement. Above all its curving up freely at the end resembles a uniqueness and free flow as in a 'signature', and also reminds of a flock of birds raising necks in sync to the sun.

Fig 61: The Structure of Time here is a combination of Kal Chakra – Time Wheel - with spokes, layers and repetitions over months and seasons and years and ages but drawn in four colors and artistic variations.

Fig 62 & Fig 63: To conclude Section 3 and to start section 4 which is Natural History, the model of seeds within seeds is used here.

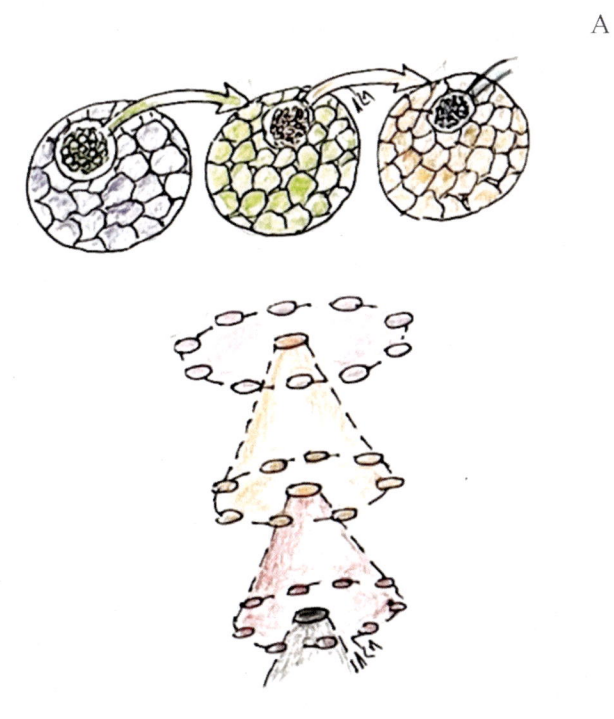

A

Nyagrodha tree seed (in some versions a Banyan tree seed) is broken by a disciple as per instructions from his teacher. He finds that there are smaller seeds in it. As he is told to open one of them, he finds still the same pattern: smaller seeds within every one of these seeds. So at every layer self-similar fractals are observed in the process till we note that there is change in scale but not change in essential content - the pattern of quadrality. The same is shown here in a front and an elevation view. This story is the famous story from Chhandogya Upanishad and explains beautifully and pertinently the unity of Atman and Brahman: Microcosm and Macrocosm.

SECTION 4

Fig 64: The beginning of the discussion on natural history ss symbolized here through a floral pattern emerging from a seed point – a common idea in Indian art - also displaying mirror symmetry around one axis.

Fig 65: An evolutionary progress from single cell to homo sapiens is seen here in a strip.

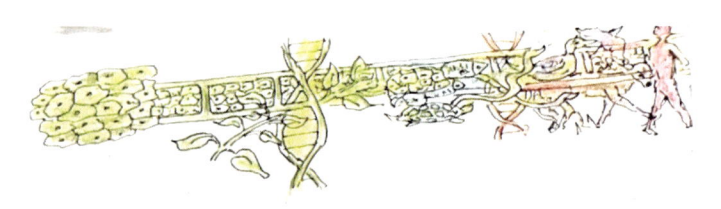

210

Fig 66: A collage of many images in section 4 and a few

from section 3 are put here as we start section 4 and see the living cell as the fulcrum of natural history.

Fig 67: There is a striking similarity in how the crystallization and solidification process happens in metals with its phase diagrams, Temperature-Time-Transformation curves in casting, welding processes, and heat treatments on one side, and cell formation, cell to organ, and then body formation in living beings. The boundaries between living and nonliving are blurred when we discuss the evolutionary paradigm with quantum certainty.

211

Fig 68: The living cell with four parts and a fourfold color scheme, continued from earlier, to depict four factors in essence.

Fig 69: The urge to live, to survive and multiply, to search for food and safeguard from predator is in each living species including insects as seen here. Also related to experiment about shampoo and lice - and testing hypothesis - exposure to shampoo actually caused mutation to the resistance to shampoo.

212

Fig 70: Shows sprouting of a seed into a plant reaching to the Sun.

Fig 71: Living cell is growing in a cluster and then into a nervous system.

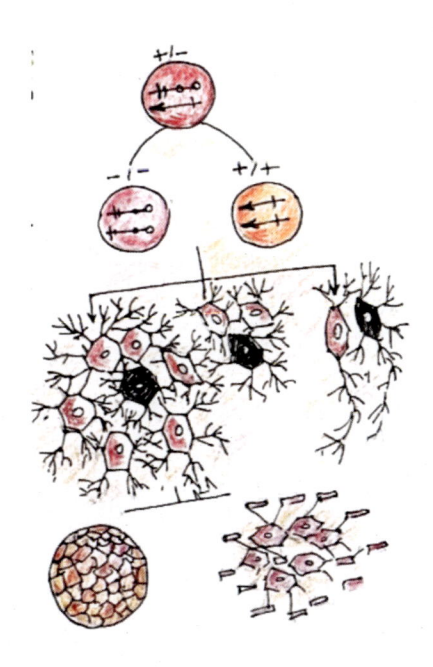

213

Fig 72: Green color signifies Life and is shown here in an intersecting pattern of lines making a whole from parts.

Fig 73: Human body with individual cell and DNA. Looking inside the cell, into the chromosome, nucleotides, and gene segment of DNA.

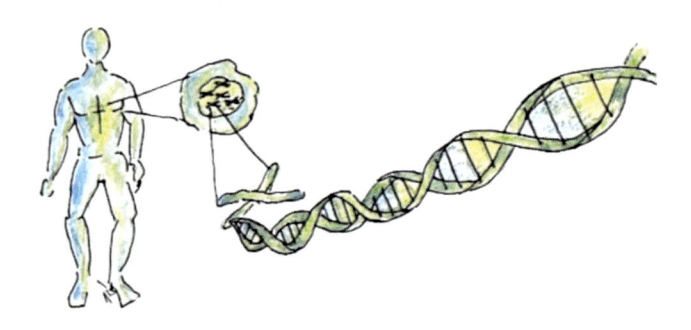

Fig 74: Selective piece from mutation and cells evolving in a full structure – zygote-mutation and mosaic. Induced mutation during embryo development in somatic cell leads to particular tissue remaining affected in adulthood.

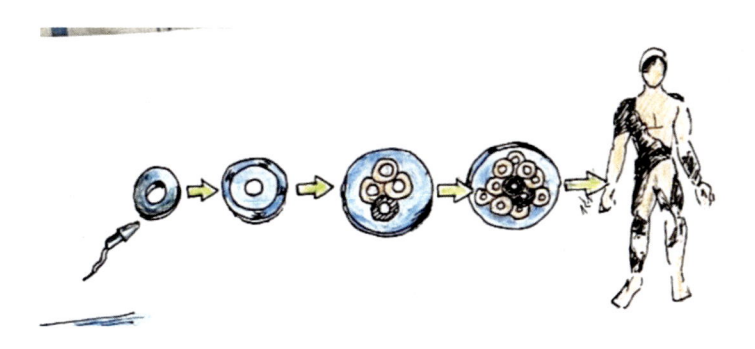

Fig 75: A branch of a tree with many smaller subbranches and leaves, and blue sky with Light behind, is leaning down and perhaps has no strength to rise up. And yet it will fall on the ground to give something for future …reminds us that Life is growth, flexibility, survival against all odds, hope, regeneration and much more.

215

Fig 76: The cells are not just chemical assemblage with formulae of eventualities. They have life themselves, have in fact mind…shown here in a dramatic way. Cells interact and their elements, like living and thinking beings, exchange, think, plan, and determine the outcomes. Note that The Mother's conversations with Satprem are published as Mother's Agenda, and Satprem has written at length about the concept of 'The Mind of Cells' which is the idea behind this visualization.

Fig 77: A collage of some of the earlier illustrations put together with the significant addition of a human figure depicting evolution of consciousness in the subtle body, and from within, having a mix of human and tree in its figure. Further, a snail spiral signifying slowing down of Light speed, a flower with symmetry in petals and an inherent spiral, efflorescence through a central flower and seeds, a polar symmetry of spirals, and addition of small birds on the spiral moving out…all signify the path traced in section 4 about natural history.

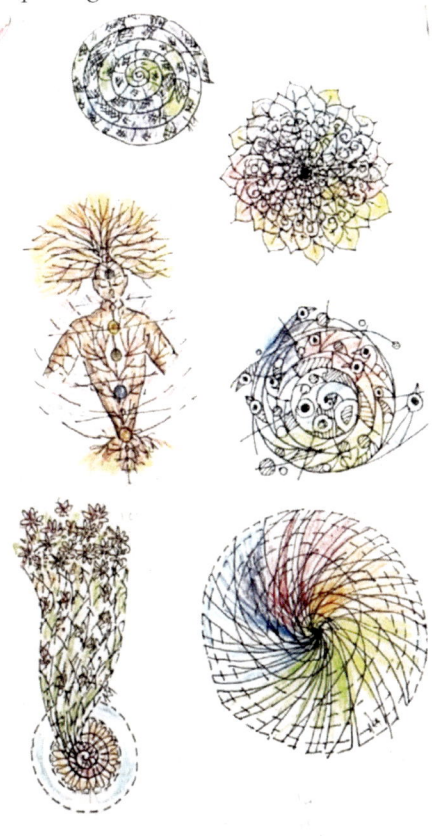

217

Fig 78: As section 4 concludes and section 5 is starting, another collage of different illustrations from the previous and next sections are put together.

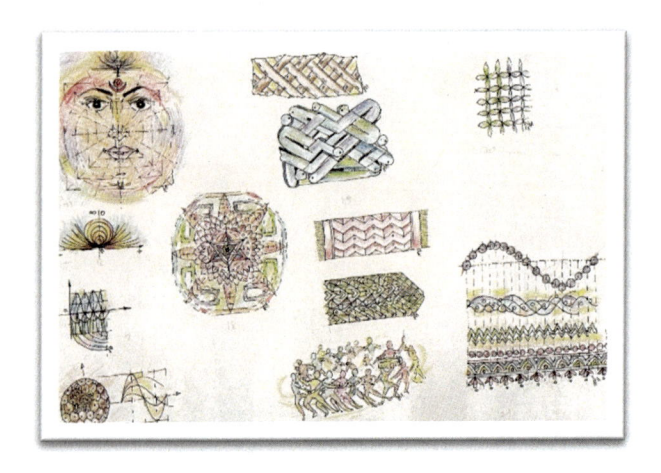

SECTION 5

Fig 79: A part of a tree with larger to smaller and still smaller leaves emerging in a pattern. Science proved that trees have feelings, urges, and sensations. The tree branch and leaves also reminds one of the Upanishadic and the Gita verses about the image with significance for a holistic science of life and evolution. Verses such as 'Shankuna Sarvani parnani' or 'Aham Vrukshasya reriva' are leading to the vision wherein the Almighty, The Pure Existence says that 'I am in every branch and every leaf and even every line on a leaf.'

Fig 80: Feelings and sensations are visualized with a symmetrically emerged beauty and tenderness of a Pushpa Flower mandala, highlighted with a soothing, faint color scheme.

219

Fig 81: As we move to sustainable human history, an iconic image of the Indian deity resembling Nataraja (Shiva in dance pose) with many arms is visualized here with some suitable adaptation and modification to suit

this subject in discussion. It is here in four colors with several heads, several hands and legs and several dance positions. With waves emerging all around it. All this is signifying the dynamic balance and equilibrium in the various upheavals throughout the human history.

Fig 82: Individuals form small aggregates, and then are formed even larger aggregates, till it comes to nations. Each entity is attempting to climb up to its objective or collective vision/objective shown here as individuals and groups doing so. This is happening while the discs and containers of the universe are in whirlwind with different motions and forces on each layer. This is creating struggle on immediate fronts, but harmony in the long run.

Fig 83: A group of human beings shown here in intense dance or interaction moving in a synchronous way as they hold some props which in a way connects them all.

Fig 84: This interrelated-ness in human aggregates and living systems or even in the apparently inanimate systems, is part of many arts and crafts of India, and the rest of the world, including and especially the tribal world. Seen here is the way bamboo strips are joined by a craftsman and with few modifications to this joining craft, we have a resemblance of several human beings interacting and exchanging energy and ideas with each other.

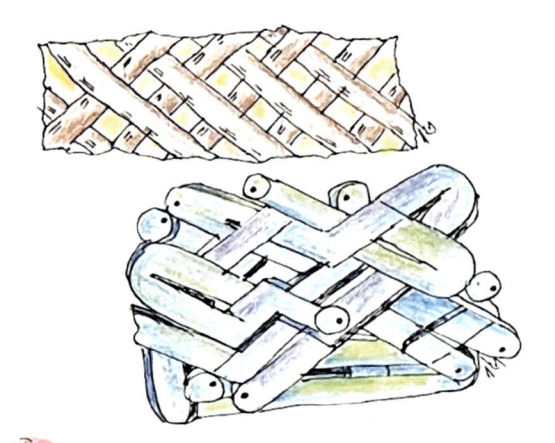

Fig 85: A sea which seemingly appears so calm and quiet and yet has the potential of up roaring, deluging everything, destroying, and ascending with giant tides. Tiny drops of water on a sea surface join together and become part of a small ascent, a small wave, which turns into larger waves and then to a giant tide, only to descend back into the sea to become one with the calm and quiet sea again. A tide ascends high in the sky as if the shining sun is attracting it. Magnify or zoom in the image to see how every part is drawn in this image.

Fig 86: Zoom in view of one part of the wave.

This is inspired by the following quote from Sri Aurobindo - "It is one Intelligence looking at itself from a hundred view-points, each point conscious of and enjoying the existence of the others. The shoreless stream of idea and thought, imagination and experience, name and form, sensation and vibration sweeps onward for ever, without beginning, without end, rising into view, sinking out of sight; through it the one Intelligence with its million self-expressions pours itself abroad, an ocean with innumerable waves. One particular self-expression may disappear into its source and continent, but that does not and cannot abolish the phenomenal universe. The One is for ever, and the Many are for ever because the One is for ever. So long as there is a sea, there will be waves." (The Three Purushas – Sri Aurobindo, Essays in Philosophy and Yoga)

Fig 87: Quantum Certainty and The Human History is visualized here through the folk dances as seen in many communities across India and the world. The group soul dominates, not the individual. There is essential oneness, connectedness, interlocks, and rhythm from the center that leads all in a cyclic to-and-fro motion, forward and backward and circles and cycles, which is resulting in harmony and sheer joy for all who perform and even see it.

Fig 88: The discussion through all sections and especially that in section 5, and moreover, the ever-inspiring thoughts from The Human cycle, The Ideal of Human Unity, (Social and Political Thought of Sri Aurobindo) along with the relevant parts from The Life Divine have created the following image and also the poem /paragraph:

"Thousand blows break the darkness, Thousand paths seek the Light, Thousand ripples create Impressions, Thousand Impressions make a mind, Thousand minds seek together, to make the whole **Life Divine**!"

The image shows, especially if zoomed in, cell- or specie- or human society- like elements, which are in turn made up of group souls and group minds, continuously and collectively aspiring to reach higher towards the supreme source of Light.

Fig 89: Zoomed in view of the interaction of various aggregates seeking thousand paths to light

Fig 90: Shows a balance where groups of people and communities are pulled up by some individuals, and possibly simultaneously some other groups of people are pulled down by other individuals. Or at times both

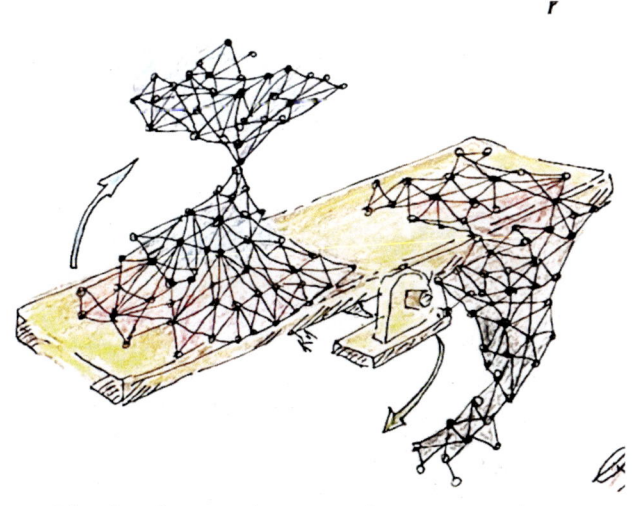

227

movements happen due to a powerful leader whose innovation is disrupting enough to pull the rest of the masses in either way. Toynbee's words, 'dominant minority should become creative minority,' can be remembered here. As Sri Aurobindo and Nolini Kant Gupta wrote, it is the nucleus which can have movement outward to bring more people in its whirl to make human societies move up and more positively move together towards the ideal of human unity. All movements start with a small group. This is analogous to the crystallization process in metals: a small nucleus is formed first, then few atoms get connected to it to make a grain. If it reaches a critical mass or a critical radius it survives, and later become the cause for further grain development and solidification of a complete body. However if before that, if the collectivity is disrupted by any external force - agitation or anything - the formation of crystals is disintegrated again and becomes one with the yet to cool down liquid in a solidification process.

Fig 91: Equation 2.8.3 reproduced in section 5, is visualized as a Space Matrix with Mandala of Light, formed from four fold symmetry and having hidden layers at the bottom and at the corners.

Fig 92: Leveraging 2.7.3 Fourfold Matrix with four sets at

four sides is seen here as a square matrix with four colors, but now a female face is superimposed on the matrix. She is the Adi Shakti, The Divine Mother, The Conscious Force.

Her Omniscient, Omnipotent, Omnipresent and Omninurturing aspects are four aspects of the Mother and have guided societies and nations throughout the history of Human Civilization.

Fig 93: On a pipal tree leaf is seen Sri Krishna playing his flute - a powerful and popular cultural idea is that at the time of pralaya, the absolute annihilation, even a pipal leaf touched and blessed by the Lord can sustain and float the world on its surface.

Connecting this to the observation here in this book, 'Here the activation-state can be further specified by leveraging (2.7.4) – The Generalized Equation for Organizational Direction…Further, and as per the discussion in the previous chapter, assume that there is a deep wanting felt by a threshold number of people. Perhaps this wanting…a step towards a global sustainable civilization.'

Fig 94: The Matrix equation here is visualized as multilayered organic pattern. Each element is a pipal leaf here as can be seen in zoomed in view.

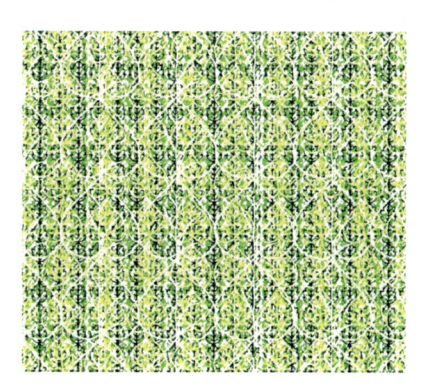

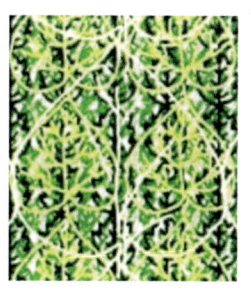

Fig 95: The Matrix created earlier is illuminated now by the Knowledge, Power, Presence and Harmony of Light. There is a sentence in the center of it: 'Tat Twam asi Shwetketu'.

231

This is one of the mahavakyas - great revelations of ancient Indian thought which is eternally true.

It is in the Chhandogya Upanishad and means:

"That Thou Art (dear pupil) Shwetketu. That which is cosmic and Infinite is also in you, in fact you are that: individual-aggregate-society-nation-universe-cosmos all are ultimately united in that One and you are that One."

This is the same story which started with the seed within the seed and fractal patterns from the micro to macro. Here is the classic concluding part of that story. The teacher finally says:

"That thou art ..Tat twam asi."

And when the disciple looks lost and wonders why then he is not able to see it, the swift answer is:

"Shraddhaya Soumya: Oh dear one, to see this - Have faith in yourself and faith in the (Qualified determinism of this) cosmos."

Fig 96: This image is inspired by the discussion in this book, and especially that in section 5.

It is originated from the great words of Sri Aurobindo from his works, esp. The Human Cycle, (Social and Political Thought), The Secret of The Veda, and The Life Divine.

In his essay 'World wars: war is a father of everything' he says:

'This Immeasurable without end or middle or beginning is he in whom all things begin and exist and end. This Godhead who embraces the worlds with his numberless arms and destroys with his million hands, whose eyes are suns and moons, has a face of blazing fire and is ever burning up the whole universe with the flame of his

233

energy. The form of him is fierce and marvelous and alone it fills all the regions and occupies the whole space between earth and heaven. ... It has enormous burning eyes; it has mouths that gape to devour, terrible with many tusks of destruction; it has faces like the fires of Death and Time. The kings and the captains and the heroes on both sides of the world-battle are hastening into its tusked and terrible jaws and some are seen B.G.11.28 |....

No real peace can be till the heart of man deserves peace; the law of Vishnu cannot prevail till the debt to Rudra is paid.'

(https://incarnateword.in/compilations/world-wars)

'For the upward movement of Brahmanaspati's formations Rudra supplies the force. He is named in the Veda the Mighty One of Heaven, but he begins his work upon the earth and gives effect to the sacrifice on the five planes of our ascent. He is the Violent One who leads the upward evolution of the conscious being.'

(The Secret of the Veda – Sri Aurobindo - https://sriaurobindo.co.in/workings/sa/10/0035_e.htm)

'And if it does not, the revolting individual flings off the yoke, declares the truth as he sees it and in doing so strikes inevitably at the root of the religious, the social, the political, momentarily perhaps even the moral order of the community as it stands, because it stands upon the authority he discredits and the convention he destroys and not upon a living truth which can be successfully opposed to his own. The champions of the old order may be right when they seek to suppress him as a destructive agency perilous to social security, political order or religious tradition; but he stands there and can no other, because to destroy is his mission, to destroy falsehood and lay bare a new foundation of truth.'

(The Human cycle – Sri Aurobindo)

Here a graph of rise and fall of civilizations on the Y axis and Time on the X axis is plotted. It is based on Fritjof Capra's 'The Turning Point' and Arnold Toynbee's writings coupled with Sri Aurobindo's Social and Political thought in his various books and mainly in The Human Cycle. In describing the Individual Age and describing world wars he compares this iconoclastic destructive god Rudra as the one clearing the ground for the next creation. If Shiva or Ashutosh is his calm and creating, maintaining part (rising of cultures and civilizations), then Rudra is the ferocious and destructive part of him (the fall of civilizations and cultures). Both are maintaining equilibrium and thus are essential elements of quantum determinism.

(The same image is on Cover page also.)

***

Relevant Background and Follow-up Information

The Author's Early Books

1. The Flowering of Management
2. India's Contribution to Management

The Fractal Series

1. Connecting Inner Power with Global Change: The Fractal Ladder
2. Redesigning the Stock Market: A Fractal Approach
3. The Flower Chronicles: A Radical Approach to Systems and Organizational Development
4. The Fractal Organization: Creating Enterprises of Tomorrow

The Cosmology of Light Series

1. A Story of Light: A Simple Exploration of the Creation and Dynamics of this Universe and Others
2. Oceans of Innovation: The Mathematical Heart of Complex Systems
3. Emergence: A Mathematical Journey from the Big Bang to Sustainable Global Civilization
4. Quantum Certainty: A Mathematics of Natural and Sustainable Human History
5. Super-Matter: Functional Richness in an Expanding Universe
6. Cosmology of Light: A Mathematical Integration of Matter, Life, History & Civilization, Universe, and Self

The Application of Cosmology of Light Series

1. The Emperor's Quantum Computer: An Alternative Light-Centered Interpretation of Quanta, Superposition, Entanglement and the Computing that Arises from it
2. The Origins and Possibilities of Genetics: A Mathematical Exploration in a Cosmology of Light
3. The Second Singularity: A Mathematical Exploration of AI-Based and Other Singularities in a Cosmology of Light
4. Triumph of Love: A Mathematical Exploration of Being, Becoming, Life, and Transhumanism in a Cosmology of Light

The Artistic Interpretation of Cosmology of Light Series

1. The Mandala Illustrated Story of Light
2. Musings on Light: A Meditative, Non-Mathematical Summary of a Cosmology of Light
3. The Illustrated Oceans of Innovation: The Mathematical Heart of Complex Systems Depicted in Indian Arts
4. Emergence Illustrated: A Mathematical Journey from the Big Bang to Sustainable Global Civilization Depicted with Indian Mythological Arts
5. The Dawn of Flame-Beings: Mythological Musings Based on a Cosmology of Light
6. Quantum Certainty Illustrated: A Mathematical Journey Through Natural and Sustainable Human History Depicted with Art

Note on Genesis of Books

In the earlier stage I wrote 'The Flowering of Management' and 'India's Contribution to Management'. The impetus for both these books was similar in that they were reactions to the environment that I was placed in at the time. When I first began working in the corporate world the reality of the environment struck me as decidedly anachronistic. I had a different sense of what life could offer and wrote 'The Flowering of Management' to capture aspects of a vision I thought corporations and money existed for. Similarly, when I wrote 'India's Contribution to Management' it was the result of the dissatisfaction I experienced when confronted with the prevalent interpretation of India. This was precipitated by my working with a US-based company, A.T. Kearney, in India. I sought to reverse that interpretation with 'India's Contribution to Management' which aimed to capture my understanding of the essence and deeper capacity of synthesis of India.

The next phase was marked by a strong interest in fractals that primarily stemmed from my beginning to see similar patterns in seemingly distinct areas of life. I wrote 'Connecting Inner Power with Global Change: The Fractal Ladder' as a theoretical framework of fractals. The fractals that I envisioned included the added complexities of emotional and thought components. This was followed by 'Redesigning the Stock Market: A Fractal Approach' which was an application of the theoretical fractal framework to the then recent global financial crises of 2008. 'The Flower Chronicles' sought to make the gist of the ubiquitous fractal I had described in the previous two books easily graspable at the visceral level primarily through many practical examples spanning diverse walks of life. 'The Fractal Organization: Creating Enterprises of Tomorrow' was a comprehensive summary of the fractal framework that included the basic theory, the applications, and a practical field guide that had been developed while I was working at the Organizational Development group at Stanford University Medical Center.

The most recent phase has focused on creating a mathematical framework to take the previously developed fractal framework further. The development of such a mathematical framework that seeks to frame innovation in complex adaptive systems was also the focus of my doctoral work. This gave birth to an exciting period and will result in multiple series of books.

The first series, comprising of six books, extended my inquiry into mathematics and complex adaptive systems to an interesting limit culminating in the nature of Light and the Cosmos. The fractal mathematics I propose is at the heart of this series: Cosmology of Light.

The first book, 'A Story of Light: A Simple Exploration of the Creation and Dynamics of this Universe and Others' contains the main ideas in the mathematics, in non-mathematical terms, that are further explored mathematically in the remaining books in this series. The second book 'Oceans of Innovation: The Mathematical Heart of Complex Systems' describes my interpretation of the mathematical foundation of complex systems. The third book, 'Emergence: A Mathematical Journey from the Big Bang to Sustainable Global Civilization' applies the mathematics to several layers of matter and life. The fourth book, 'Quantum Certainty: A Mathematics of Natural and Sustainable Human History' describes a process culminating in space, time, energy, and gravity quantization by which history is made. The fifth book, 'Super-Matter: Functional Richness in an Expanding Universe' describes a process for the creation of super-matter-based on a need for continued functional-richness. A link is made between the resulting quantization of space and an expanding universe. The final book, 'Cosmology of Light: A Mathematical Integration of Matter, Life, History & Civilization, Universe, and Self' proposes an integrated mathematical framework that flows through all things, hence unifying matter, light, civilization, history, universe, and self.

The second series further explores the implications of "one mathematics flowing through all things". The first book in this series 'The Emperor's Quantum Computer: An Alternative Light-Centered Interpretation of Quanta, Superposition, Entanglement and the Computing that Arises from it' describes an alternative narrative for quantum computing to the one commonly expressed today. The second book in the series, 'The Origin and Possibilities of Genetics: A Mathematical Exploration in a Cosmology of Light' explores pre-genetic, genetic, and post-genetic possibilities in a cosmology of light. This book, 'The Second Singularity: A Mathematical Exploration of AI-Based and Other Singularities in a Cosmology of Light' explores the limits of AI-based singularities with respect to light-based singularities. This book, the final in this series explores transhumanism in a cosmology of light.

The third series, Artistic Interpretation of Cosmology of Light, is intended to make Cosmology of Light more accessible by interpreting it artistically. The first book in the series is 'The Mandala Illustrated Story of Light'. The objective is to lead the reader through the story of light using mandalas as an aid in the journey. The second book 'Musings on Light' is a meditative book set to graphical illustrations. The illustrations focus on 50 key concepts derived from the ten-book joint Cosmology of Light series. The third book 'The Illustrated Oceans of Innovation: The Mathematical Heart of Complex Systems Depicted in Indian Arts' uses Indian Arts to illustrate the mathematical heart of complex systems. The fourth book

'Emergence Illustrated: A Mathematical Journey from the Big Bang to Sustainable Global Civilization Depicted with Indian Mythological Arts' uses Indian mythological arts to express the concepts of Emergence. The fifth is book 'The Dawn of Flame-Beings: Mythological Musings Based on a Cosmology of Light' that uses illustrations to help focus on the mythological aspects of a cosmology of light. The sixth and current book is 'Quantum Certainty Illustrated: A Mathematical Journey of Natural and Sustainable Human History Depicted with Art' that leverages pencil drawings and other art to shed insight into the mathematical development of history. Several additional books are envisioned and are currently under development.

About the Author

Dr. Pravir Malik has been developing a unified theory and mathematics of organization over the last three decades. He has written 20 books related to this to emphasize a whole systems approach integrating individual, organizational, economic, social, environmental, and evolutionary dimensions. In recent years he has been intimately involved with computer modeling of complex organizational, economic, and world systems to help different stakeholders practically navigate and understand possible futures. He is also a regular contributor to Forbes and recently completed a ten-part series on the creation of sustainable wealth through interpreting environments as complex adaptive systems. In 2020 he also designed and delivered a multi-part Organizational Sciences Certification program with Forbes that was attended by executives from 250 companies. This program included a path-breaking approach to bringing about organizational change by leveraging light and was derived from Dr. Malik's 10-book series on Cosmology of Light.

Dr. Malik has held a number of leadership positions. He is currently Chief Strategy Officer at Galaxies, focused on modeling of complex realities, and Chief Technologist at Deep Order Technologies where he is spearheading the development of a revolutionary atom-based quantum computer. Dr. Malik has deep interest in core technologies such as quantum computing, artificial intelligence, genetics, and transhumanism. His recent writings propose alternative trajectories of development for each of these areas to increase the likelihood of sustainable global development. Formerly he has been Head of Organizational Sciences at Zappos.com - an eCommerce company, Managing Director of BSR - a global sustainability and environmental consulting company, and a member of the Founding Team of A.T. Kearney India - a global operations-focused management consulting company.

He has a Ph.D. in Technology Management with a focus on Mathematics of Innovation in Complex Adaptive Systems from the University of Pretoria, an MBA from Northwestern University's J.L. Kellogg Graduate School of Management with a focus on Marketing and Organizational Behavior, an MS in Computer Science from the University of Florida with a focus on AI, and a BSE in Computer Engineering from the Case Western Reserve University. He has also served on the faculty of Sri Aurobindo International Center of Education at the undergraduate level. Pravir is a global citizen who has lived, worked, and been educated in many parts of the world.

About the Illustrator

Dr. Narendra has three decades of extensive experience in academics, industry, projects, grassroot work on social and cultural issues, editing magazines and journal & his recent passion has been in multidisciplinary research and its artistic expressions.

He has completed his graduation and post graduation in Production engineering from University of Mumbai and ranked first in his Master's program. He completed several technology, management and entrepreneurship certifications from reputed institutes like IIT Bombay, IIM Ahmedabad, ISB Hyderabad, NEN-STVP and IIM B. He completed his Ph.D. from Hindu University of America. His thesis was on an unusual subject: 'A study of philosophy and Futurology of Artificial Intelligence in the light of Sri Aurobindo's integral thought.'

Inspired by the message of Swami Vivekananda and Sri Aurobindo for future of India and humanity and intrigued by the shift in foundations of Physics and the future shock, he switched over from his industrial career and worked as full-time dedicated worker for a leading NGO, Vivekananda Kendra, Kanyakumari and was posted in Northeast India in various remote areas of Assam and Arunachal Pradesh after extensive training in Kanyakumari for one year. Perplexed by the complex socioeconomic developmental especially cultural and human problems there, he started studying works of Sri Aurobindo more deeply. He was given a research project which was later published as a book named 'Ashwattha' by Vivekananda Kendra. This book has hundreds of illustrations by him and since then he has written many papers for conferences, journals, and magazines on diverse topics from technology, whole brain thinking, creativity and innovation, management, Indian arts, consciousness studies, history, evolution of consciousness, appropriate technology and sustainability, science and technology in Ancient India, Indian ethos in management, Indian philosophy, psychology and Sri Aurobindo studies. His paper on Indian ethos in management won the award from BMA for best paper for the year. He has served in industry and in academics holding various key positions including principal, project head, academic consultant, etc. He was benefitted in this journey by association and guidance from many experts.

Dr. Narendra has worked as Principal for eight years and faculty for nine years in engineering fields and he is proactive in assessing training & development needs and effectively aligning programs / interventions with business /social objectives. He is deft in designing innovative programs for industry as well as academics while catering to emerging industry needs. He has a distinguished inclination to social responsibility exemplified in interdisciplinary research projects, field work and development of innovative projects dedicated to integral and futuristic studies and also for discovering the underlying currents of culture and psychology in it. At present he is working as Project Director for Vivekananda Prabodhini, Mumbai. He also provides Academic consultancy especially for distant, digital and open university models wherein students from underprivileged and remote areas have urge to learn further.

However more than all these things, he is an artist by heart and has developed passion for sketching, painting in traditional as well as digital art over the years. His sketches are now part of several magazines, blogs, books and forums. He has found that illustrating a text helps in deeper thought expression and provides further insights especially for path breaking and transdisciplinary subjects such as this book.

Selected Author Online Presence

- Amazon Author Page: https://www.amazon.com/Pravir-Malik/e/B002JVAEZE
- LinkedIn Profile: https://www.linkedin.com/in/pravirmalik/
- Forbes Page & Articles: https://www.forbes.com/sites/forbeshumanresourcescouncil/people/pravirmalik1/#1fa1097c17be
- Forbes Profile: https://profiles.forbes.com/members/hr/profile/Pravir-Malik-Head-Organizational-Sciences-Zappos/44463250-f2ab-434a-b1e2-0a6bdf54d970
- Google Scholar Page: https://scholar.google.com/citations?user=7DWWWZ8AAAAJ&hl=en
- Sage Author Page: https://us.sagepub.com/en-us/nam/author/pravir-malik
- IEEE Profile: https://ieeexplore.ieee.org/author/37086022058
- YouTube Page: https://www.youtube.com/user/Aurosoorya
- Twitter: https://twitter.com/PravirMalik
- Research Gate Profile: https://www.researchgate.net/profile/Pravir_Malik

- Eventbrite: https://www.eventbrite.com/o/pravir-malik-30159112262
- Medium: https://medium.com/@PravirMalik
- Company website: http://www.deepordertechnologies.com/

REFERENCES

1. Al-Khalili, Jim. 2014. Life on the Edge: The Coming of Age of Quantum Biology. Bantam Press. London
2. Arkani-Hamed, Nima; Bourjaily, Jacob L.; Cachazo, Freddy; Goncharov, Alexander B.; Postnikov, Alexander; Trnka, Jaroslav (2012). "Scattering Amplitudes and the Positive Grassmannian". arXiv:1212.5605
3. Bar-Yam, Y. 2016. From big data to important information. Complexity, 21: 73–98. doi:10.1002/cplx.21785
4. Brumfiel, G. 2008. Physicists spooked by faster-than-light information transfer. Nature.
5. Chown, M. 1990. Can Photons Travel Faster Than Light? New Scientist 126(1711)
6. Clegg, Brian. 2014. The Quantum Age: How the physics of the very small has transformed our lives. Icon Books: London, UK.
7. De Broglie, L. 1929. The Wave Nature of the Electron. Nobel Lecture. http://www.nobelprize.org/nobel_prizes/physics/laureates/1929/broglie-lecture.pdf
8. Deep Order Mathematics Videos. 2016. Deep Order Technologies. http://www.deepordertechnologies.com/new-index
9. DeZurko, E.R., 1952. Greenough's Theory of Beauty in Architecture. *Rice Institute Pamphlet-Rice University Studies, 39*(3).
10. Diamond, J. 2005. Collapse: How Societies Choose to Fail or Succeed. Viking Books: New York
11. Ebeling, W, Sokolov, I. 2005. Statistical Thermodynamics and Stochastic Theory of Nonequilibrium Systems. Singapore: World Scientific Publishing
12. Einstein, A. 1995. Relativity: The Special and General Theory. New York: Broadway Books.

13. Feinberg, G. 1970. "Particles That Go Faster Than Light". Scientific American. Feb 1970, 69-77.
14. Feynman, RP. 1985. QED The Strange Theory of Light and Matter. New Jersey: Princeton University Press
15. Goodsell, David. 2010. The Machinery of Life. New York: Springer
16. Gottlieb, M. 2013. The Feynman Lectures on Physics: III. California Institute of Technology. http://www.feynmanlectures.caltech.edu/III_1 6.html
17. Gubser, S. 2010. The Little Book of String Theory. Princeton University Press
18. Hawking, Stephen. 1988. A Brief History of Time. New York: Bantam Books
19. Isaacson, W. 2008. Einstein: His Life and Universe. Simon and Schuster. New York.
20. Jeans, J. 1932. The Mysterious Universe. Cambridge University Press.
21. Johnson, S. 2010. Where Good Ideas Come From: The Natural History of Innovation. Riverhead Books: New York
22. Kaufmann, S. 1995. At Home in the Universe. New York: Oxford University Press.
23. Laszlo, E. 2014. The Self-Actualizing Cosmos: The Akasha Revolution in Science and Human Consciousness. Inner Traditions: Vermont
24. Lorentz, H.A. 1925. The Science of Nature. Vol. 25, p 1008. Springer
25. Malik, P. 2009. Connecting Inner Power with Global Change: The Fractal Ladder. New Delhi: Sage Publications
26. Mora, C, Tittensor, D, Adl, S, Simpson, A, Worm, B. 2011. How Many Species Are There on Earth and in the Ocean? Plos.org. http://journals.plos.org/plosbiology/article?id=10.1371/journal.pbio.1001127#abstract0
27. Ogburn, W. Tomas, D. 1922. Are Inventions Inevitable? A Note on Social Evolution. Political Science Quarterly 37, no.1

28. Pauli, W. 1964. Nobel Lectures, Physics 1942 – 1962. Elsevier Publishing Company. Amsterdam.
29. Pearson, R. 1997. "Frontier Perspectives", Journal of the Center for Frontier Sciences at Temple University. Spring/Summer 1997, Volume 6, Number 2, ISBN: 1062-4767.
30. Perkowitz, S. 2011. Slow Light. London: Imperial College Press
31. Planck, M. 1933. Where is Science Going? Ox Bow Press. Connecticut.
32. Portugali, J., 2012. *Self-organization and the city*. New York: Springer Science & Business Media.
33. Prigogine, I. 1977. Time, Structure, and Fluctuations. *Nobelprize.org*. Nobel Media AB 2014. Web. 5 Mar 2016. <http://www.nobelprize.org/nobel_prizes/chemistry/laureates/1977/prigogine-lecture.html>
34. Saunders, S, Barrett, J, Kent, A, Wallace, D. 2012. Many Worlds? Everett, Quantum Theory and Reality. Oxford: Oxford University Press
35. Schrodinger, Erwin. 1944. What is Life? Cambridge University Press. Cambridge.
36. Schrodinger, E. 1995. The Interpretation of Quantum Mechanics. Ox Bow Press. Connecticut.
37. Salkind, N. 2007. Encyclopedia of Measurement and Statistics. Thousand Oaks: Sage Publications.
38. Snyder, M. 2010. Stanford Medicine. http://med.stanford.edu/news/all-news/2010/03/what-makes-you-unique-not-genes-so-much-as-surrounding-sequences-says-stanford-study.html#.html
39. Sri Aurobindo. 1971. Social and Political Thought. Sri Aurobindo Ashram Press: Pondicherry
40. Stewart, Ian. 2012. In Pursuit of the Unknown. Basic Books. New York.
41. Taleb, N. 2010. The Black Swan: The Impact of the Highly Improbable. Random House Paperbacks: New York

42. Thomas, D, Brown, J. 2009. Why Virtual Worlds Can Matter. International Journal of Learning and Media 2009 1:1, 37-49
43. Toynbee, A. 1961. A Study of History, Volumes I – XII. Oxford University Press: Oxford
44. Usó -Doménech, J, Nescolarde-Selva, J, Lloret-Climent, M, Fan, M. 2016. Semiotic open complex systems: Processes and behaviors. Complexity, 21: 388–396. doi:10.1002/cplx.21817
45. Vivoli, C. 2016. The Experiment that will allow Humans to "See" Quantum Entanglement. MIT Technology Review: Cambridge, MA.
46. Weizenbaum, J. 1976. Computer Power and Human Reason: From Judgment to Calculation. San Francisco: W.H. Freeman
47. Whitaker, A._2006. Einstein, Bohr and the Quantum Dilemma: From Quantum Theory to Quantum Information. Cambridge: Cambridge University Press
48. Wilczek, F. 2016. A Beautiful Question: Finding Nature's Deep Design. New York: Penguin Books
49. Wolchover, N. 2013. A Jewel at the Heart of Quantum Physics. Quanta Magazine. https://www.quantamagazine.org/20130917-a-jewel-at-the-heart-of-quantum-physics/
50. Wright, R. 2009. Evolution of Compassion. https://www.ted.com/talks/robert_wright_the_evolution_of_compassion/transcript?language=en. TED 2009.
51. Yates, F.E. 2012. *Self-organizing systems: The emergence of order*. New York: Springer Science & Business Media.

Made in the USA
Columbia, SC
12 July 2023

45f88d19-f205-4333-9175-5331284f034eR01